Greater Manchester Buses

Greater Manchester Buses

Stewart J. Brown

Capital Transport

First published 1995

ISBN 185414 174 0

Published by Capital Transport Publishing
38 Long Elmes, Harrow Weald, Middlesex

Printed in England

© Stewart J. Brown 1995

Before the turbulent times of deregulation, GMT was contemplating what the next generation of buses might be like. In 1985 Ken Mortimer produced styling sketches including this imaginative proposal which offered good kerbside vision.

Contents

Author's Note	6
Introduction: The Great Days of Integration	9
The Formation of Selnec	10
1970: Selnec sets Sail	16
1971: Restructuring	28
1972: North Western Takeover	34
1973: Reorganised Again	46
1974: GMT Takes Over	54
1975: Rail Reversal	62
1976: LUT Takeover	68
1977: New Deliveries Pick Up	72
1978: Integration of LUT	76
1979: Late Buses and Service Cuts	82
1980: Rising Fares and Falling Patronage	86
1981: Restructuring and Cutbacks	94
1982: The Last of the Half-Cabs	102
1983: Olympian Arrival	106
1984: Light Rail Plans	110
1985: The Domino Effect	114
1986: Deregulation – The End of an Era	118
The Late 1980s: A Competitive Market	126
The 1990s: Driving Towards the Private Sector	134
Fleet List	140
Appendix 1: The Fleet at Privatisation	152
Appendix 2: Key PTE Statistics, 1970-1986	152

Author's Note

If I hadn't been born a Glaswegian, I would have wanted to be a Mancunian. I unashamedly enjoy spending time in Manchester and its environs. So writing what amounts to a history of the last quarter-century of its public transport is a pleasurable task, even if tinged by sadness at watching the fine ideals of co-ordination being systematically destroyed by political dogma at a time when it could be argued that they are needed more than ever before as congestion and pollution threaten the quality of city life.

But I come to praise Manchester's transport, not to bury it. Well, praise is perhaps over-stating the case, but Selnec and GMT did a lot to raise standards for travellers and employees and I hope I've recorded most of that in the pages which follow. It made a few mistakes too, but what organisation hasn't?

Completing this in the spring of 1995 it is hard not to feel that we will never see the likes of the Selnec/Greater Manchester PTE again – a unified organisation dedicated to the provision and co-ordination of high-quality competitively-priced public transport. We seem to have gone too far down the road of private sector profit and free-market competition to ever return.

So, I hope this volume captures something of what is already a bygone age, and some of the excitement and enthusiasm which went in to developing Britain's biggest urban bus operation outside London.

On a question of style, those who can remember the Selnec period may recollect that the PTE insisted on the use of capitals for the SELNEC name. It appeared thus in all contemporary documentation, but for ease of reading it is rendered throughout this volume as Selnec.

No work of this size can be completed without some assistance. It gives me great pleasure to record the unstinting help received from Anthony Wyer of the Selnec Preservation Society and from the many members of the Greater Manchester Transport Society who work very hard to maintain the excellent Manchester Museum of Transport – which is always well worth a visit if you're in the area. Michael Eyre, who with Chris Heaps, wrote *The Manchester Bus* (TPC) which covers the period up to the creation of Selnec, took a lot of time and trouble to comment on the text.

A large number of other individuals offered help and provided information including John Aldridge, Chris Bowles, Richard Cochrane, Darren Couperthwaite, Peter Crichton, Mark Green, Geoffrey Harding, Jim Hulme, T J Johnson, Ian Longworth, Andrew Nolan, John H A Pollock and Jonathan Rowse. Selecting photographs from the vast number available was a mammoth task. The work of individual photographers is acknowledged as appropriate, but special mention must be made of Michael Fowler, Roy Marshall and Reg Wilson who made large numbers of prints available from their extensive collections.

The Greater Manchester PTE does, of course, live on, as the authority responsible for route tendering and service promotion amongst other things. And the PTE's bus-operating legacy, not least in the shape of the standard Northern Counties double-decker, will be visible on the streets of Greater Manchester for quite some years to come.

Framilode, May 1995　　　　　　　　　　　　　　　　　　　　　　　　　Stewart J Brown, MCIT

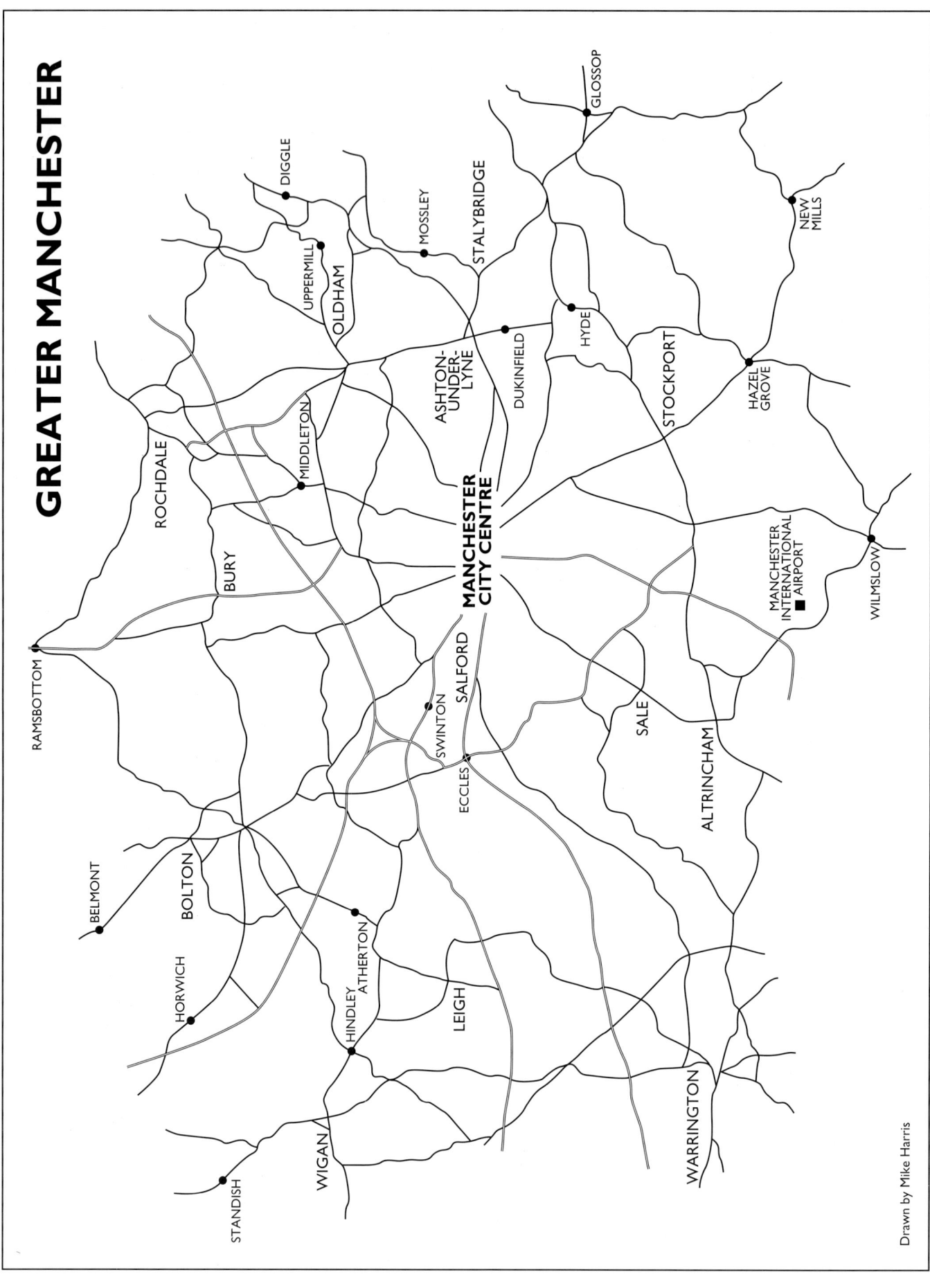

Introduction: The Great Days of Integration

The South East Lancashire and North East Cheshire Passenger Transport Authority (PTA) and its bus-operating Passenger Transport Executive (PTE) were children of the 1968 Transport Act.

Concern about urban congestion and the movement of people in towns and cities led to the creation of PTAs and PTEs in the four biggest urban areas of England at the end of the 1960s — Merseyside, Tyneside, West Midlands and Greater Manchester. The description of Greater Manchester as South East Lancashire and North East Cheshire may have been as much to respect the sensibilities and civic pride of the good citizens of the other substantial towns in the area as a wordy search for geographical precision.

The PTA comprised representatives appointed by the local authorities in the area, usually local councillors. The PTE was managed by full-time professionals headed by a director general who was appointed by the PTA. The creation of a PTA and a PTE established the separation of policy issues (handled by the PTA) from management matters (the province of the PTE) which was a significant change from the previous system whereby the local authorities both owned the bus fleets and set policy.

The Selnec PTA came into being on 1st April 1969 and the PTE on 1st September. Two months after its formation, on 1st November 1969, the Selnec PTE took over the operations of the 11 municipal bus undertakings in its area and with them 10,336 employees and a fleet of 2,514 operational motorbuses on AEC, Albion, Atkinson, Bedford, Bristol, Daimler, Dennis, Guy and Leyland chassis. This made it the biggest bus operator in the UK outside London.

Manchester Corporation's fleet accounted for just under half the total. The original Selnec constituents, in descending order of size, were:

Manchester	1,220
Salford	271
Bolton	249
Oldham	185
Stockport	148
Rochdale	129
Bury	96
SHMD	88
Ashton	59
Leigh	57
Ramsbottom	12

There was a high degree of co-ordination in the area prior to the creation of the PTA, with interworking between the municipal transport operators which in many cases dated back to joint tramway operations early in the century.

From time to time some of the municipalities had examined the scope for more formal integration of services. This thinking was boosted by the 1930 Road Traffic Act which introduced a unified system for licensing of bus services by area Traffic Commissioners, a matter previously dealt with by local authorities. In 1931, at the instigation of Oldham, representatives from Manchester, Salford, Bolton, Bury, Oldham, Stockport, Rochdale, SHMD, Ashton, Leigh and Wigan met to discuss the formation of a regional municipal transport board for South East Lancashire.

Wigan and Leigh soon pulled out of the discussions on the grounds that their services were not really tied in with the Manchester area operations, but negotiations continued during 1934 with Manchester City Council giving its approval to the principle of a joint municipal transport board.

The idea staggered on, with various towns withdrawing as they developed fears that a joint board might not really be in their best interests. By 1936 only Manchester, Salford, Oldham and Rochdale were still interested. There was a final display of interest in 1937 when all of the original 1932 municipals held a conference on the subject, along with company bus operators Lancashire United, North Western and Ribble.

But with a high degree of co-ordination in existence, underpinned by the bus licensing system set up by the 1930 Act, there was no real stimulus for municipalities to relinquish control of their operations. Travellers in the Greater Manchester conurbation already enjoyed a co-ordinated service. There was no urgent need for any change in existing forms of ownership and control.

And so it continued. Tramway systems were gradually abandoned, with the last tram in the area running in Stockport in August 1951. Trolleybuses came — and went. Co-ordination was maintained.

Above **Salford's Victoria bus station in April 1970. A 1964 Metro-Cammell-bodied Leyland Titan PD2/40 with synchromesh gearbox and vacuum brakes pulls out with no visible sign of the change which has taken place in its ownership. The two PD2s on the right carry Selnec fleet numbers.**
R L Wilson

Renumbering was a slow and sometimes haphazard process as two consecutively-numbered AEC Regents in the Rochdale fleet illustrate in May 1970. The bus on the left became 6191. R L Wilson

A 1957 Leyland Titan with Crossley body acquired from the Oldham fleet. The municipal crest has gone and the Selnec fleet number has been applied. This was one of five similar buses supplied to Oldham in 1957, but the only one to survive in the operational fleet when Selnec took over. It was withdrawn in 1973. R L Wilson

The Formation of Selnec

The idea of forming PTAs as a means of addressing traffic problems in major urban areas first received official recognition in the summer of 1966 when Harold Wilson's Labour government published a transport policy white paper in which it suggested the formation of CTAs – Conurbation Transport Authorities. The CTAs would have been involved in highways and parking, as well as running all local bus services.

The PTAs which eventually appeared had somewhat narrower powers with no involvement in highway and parking policies, and no power to take over bus services which were run by company operators. They had instead to "make agreements" with the companies. This they did, with varying degrees of success.

The Selnec PTA covered an area of 545 square miles with a population of 2.6 million people. To serve this large area the PTE, under director general Tony Harrison, was set up with three operating divisions – Central, Southern and Northern.

The Central division comprised the former Manchester and Salford operations and had a fleet of 1,491 buses of which 1,403 were double-deckers. Salford had two depots – Frederick Road and Weaste – while Manchester had seven at Birchfields Road, Hyde Road, Northenden, Parrs Wood, Princess Road, Queens Road and Rochdale Road. The Central division's headquarters were at the modern Manchester City Transport offices in Devonshire Street North, alongside Hyde Road depot.

The Northern division encompassed Bolton, Bury, Leigh, Ramsbottom and Rochdale and had 543 buses. There were three depots in Bolton and one in each of the other towns. Its headquarters were in the former Bolton Transport offices in Bradshawgate.

The smallest division, Southern, had 480 buses and incorporated Ashton, Oldham, SHMD and Stockport. Stockport had two garages; each of the other constituents had one. The divisional head office was in Daw Bank, Stockport.

Harrison was not a busman – he joined the PTE from Bolton where he had been the town's chief executive. Three other directors constituted the original PTE board. David Graham was director of finance and administration, Geoffrey Harding was director of operations and engineering, and Ernest Armstrong was director of establishment and industrial relations.

The three other PTEs formed at the same time as Selnec adopted a livery derived from

Left Short-lived acquisitions by the PTE included eight Leyland Titans in the Oldham fleet which had been new to Sheffield. They had been purchased to alleviate a shortage of serviceable buses in Oldham in 1965-66. None received Selnec livery. This example, a 1952 Leyland-bodied PD2/10, was withdrawn in 1971.
Gavin Booth

Below Selnec was eager to promote a modern image and in the early days a number of Mancunians were repainted to do just that. Park Royal-bodied Atlantean 1003 glows in Manchester's Piccadilly bus station in October 1970. The coin-in-the-slot symbols on either side of the destination screen indicate a bus equipped with a fare-box, which is visible on the driver's left. An older ex-Manchester bus, a Leyland PD2, loads in the background. R L Wilson

that used by the biggest constituent. Tyneside used Newcastle City Transport's yellow, West Midlands used Birmingham City Transport blue and Merseyside used Liverpool City Transport green – although a blue livery was used in its Wirral division for a time in deference to its customers in Birkenhead and Wallasey. The spread of the livery of the dominant operator in each of the other PTE fleets gave the impression of a takeover and this was something which Selnec was careful to avoid. As part of the search for a new identity a consumer clinic was held using painted model buses, all in colours which avoided those used by existing operators in the area – which must have seriously restricted the choices available.

Selnec's new livery, unveiled on 11th March 1970, was bright Sunglow orange and Mancunian white, colours not used by any of the operators taken over – or indeed by any other British bus operator. The livery layout was new too and at the launch stress was laid on the need to apply the colours in the correct proportions. The orange band above the lower deck windows was not to be more than 12 inches deep and the white between-decks panel above this was not to be more than 26 inches deep. This latter dimension could be maintained by varying the depth of the orange area around the upper deck windows. Although called Mancunian white, the white was a warmer shade than that which had been used by Manchester City Transport for its Mancunians.

The original layout was to have been a reversal of the style used on the Manchester City Transport Mancunians with orange for the skirt and for the area from the top of the lower deck windows to the top of the upper deck windows. The layout which was eventually chosen was similar to that used by Southport Corporation.

A stylised 'S' logo – the lazy S – was applied to the sides of the buses and incorporated the divisional fleetname. On Central buses the logo was lagoon blue, on Northern buses it was rose magenta and on Southern buses it was spring green. Coaches carried an orange logo with the fleetname Selnec Express, although this was later changed to just Selnec. In each case the fleetname was black. The lettering was in a new style created by Selnec's advertising agency, Brunning Manchester, and known as Selnec Alpha.

The new corporate identity was also applied to the Manchester City Transport parcels operation, a van fleet providing deliveries in and around the city, which became Selnec Parcels.

Renumbering the fleet

Having taken over 11 municipal fleets there was an urgent need to institute a common fleet numbering system. With most of the acquired operators having used simple sequential fleet numbers there were in some instances as many as six buses from different fleets with the same number.

Thus in March 1970 a Selnec fleet numbering system was devised which carefully and sensibly avoided the need to renumber the ex-Manchester fleet, apart from its 64 single-deck buses. This involved a deal with Salford which allowed Salford to keep its service numbers, but that was a small price to pay for having to renumber only around 350 buses or just over 20 per cent of Central's fleet.

Manchester's numbering system dated back to just after World War II when blocks of 1000 numbers were allocated to trolley-buses (1xxx), Crossleys (2xxx), Leylands (3xxx) and Daimlers (4xxx). The 1xxx and 2xxx series were revived in 1968 for the fleet's new Mancunian-style Atlanteans and Fleetlines respectively.

(1) Central division block: 1 – 4999

The ex-Salford fleet, now part of the Central division, was renumbered in what had been the Manchester series. Thus Salford's Daimler double-deckers became 4000-4090, fitting in below the lowest surviving Manchester Daimler, 4111. Similarly Salford's Leylands were renumbered 3000-3169, below the oldest ex-Manchester Leyland still in use, 3200. The Salford renumbering, with buses in blocks of consecutive numbers, differed from that of the other divisions where the new Selnec fleet number generally bore an obvious relation to the old municipal one. Manchester's single-deck bus fleet was renumbered 1-64, and Salford's nine single-deck buses became 65-73. Salford's only coach became 215, joining the end of the Manchester coach series, 201-214. Selnec reserved numbers up to 299 for coaches.

(2) Southern division block: 5000 – 5999

The highest numbered bus in the Manchester fleet was 4760 and the Southern division fleet was numbered after it in the range 5000 to 5999 with separate groups of numbers for each of the acquired fleets and for single-deckers. Numbers in the 5000-5100 range were single-deckers from each of the four fleets absorbed.

Double-deckers acquired from Oldham were numbered in the series 5101 to 5378, allocated by adding 5000 to Oldham fleet numbers below 300, and 4900 to fleet numbers above 300.

The ex-Ashton double-deck fleet was numbered in the range 5401 to 5469, generally by adding 5400 to the existing Ashton number. Similarly, former SHMD double-deck buses had 5600 added to their numbers, bringing the fleet into the 5601 to 5699 range.

Ex-Stockport double-deckers were numbered from 5801, with 5800 being added to low fleet numbers and 5620 or 5600 being added to high fleet numbers.

(3) Northern division block: 6000-6999

The Northern division fleet was numbered between 6000 and 6999. As with Southern, the first 101 numbers between 6000 and 6100 were used for single-deckers from all five acquired undertakings. Former Rochdale double-deckers became 6138 to 6244 by adding 5900 to existing fleet numbers. Those from Bury were numbered between 6302 and 6393 – most simply had 6200 added to their existing fleet numbers but some were raised by 6100 or 6390.

Ramsbottom's small fleet of double-deckers had been numbered from 1 to 10; under Selnec ownership this became 6401 to 6410. Ex-Bolton double-deckers had 6500 added to their numbers, bringing them in to the range 6563 to 6801. The Leigh fleet had been numbered haphazardly as new buses simply filled gaps in the existing series and, surprisingly, Selnec did not regroup the ex-Leigh buses in any semblance of order. The ex-Leigh double-deckers were fitted into their new number series, in the 6903 to 6965 range, by the simple addition of 6900 to existing numbers.

At one stage the Northern fleet was to have been numbered in the range 7000 to 8999 but it was decided to reserve numbers above 7000 for a new standard bus which was hinted at in March 1970 when the new livery was launched.

Above **Dual-door buses were the height of fashion in 1969 and Bury took delivery of its first in the shape of three East Lancs-bodied Atlanteans. They were Bury's last new buses. The central location of the numeral '1' shows that this bus was renumbered by the simple addition of transfers '639'. It is seen in the town centre in May 1970.** R L Wilson

Top left **In the same year that 1950 Titans were being withdrawn, so also were 1965 Panther Cubs. Manchester bought 20 Panther Cubs in 1964-65 – a decision which it soon regretted and none were recertified when their initial seven-year Certificates of Fitness expired. Panther Cub 35, with 43-seat Park Royal body, is seen near the end of its life in Manchester's Oxford Road in September 1970. Although designed for one-man-operation it is in fact carrying a conductor as the slip board below the windscreen advises. The imposing Refuge Assurance building in the background was later to become a hotel.**
R Marshall

Centre left **The oldest ex-Salford buses to appear in Selnec colours were 1962 Daimler CVG6s with Metro-Cammell Orion bodies. 4018, with Manchester-style fleet number transfers, heads along Salford's Chapel Street past the Whitbread brewery in September 1970. It was withdrawn in 1973, although a few of its type survived until 1975. Most ex-Salford buses and many ex-Manchester buses had centrally-located fleet numbers above the driver's cab. Fleet number location was one item which initially escaped Selnec's corporate design team.** R Marshall

Bottom left **The last bus delivered to Ashton was this Atlantean with Northern Counties body which became 5461 in the Selnec Southern fleet although this was not yet apparent when this photograph was taken in July 1970. It was one of five, the first buses in the Ashton fleet of two-door layout. Note the reflective registration plates only recently legalised and at this time still relatively uncommon.** G R Mills

Right **Leigh had seven single-deckers, the oldest of which were a pair of 1960 Leyland Tiger Cubs which were used as one-man buses. They had 43-seat East Lancs bodies finished in a livery style which was outdated when the buses were new, although it sits quite comfortably on the conservatively-styled body. 6001 loads in Leigh bus station in 1971; it was withdrawn in 1973.**
R Marshall

Above **The Leigh fleet was renumbered using Manchester-style transfers, as illustrated by 1957 PD2 Titan 6954. All of Leigh's double-deckers were of lowbridge or lowheight design. Not only were there numerous low bridges in the area – the garage roof was also too low to house full-height double-deckers. Like all post-1950 Leigh buses, this Leyland has an East Lancs body. A 1964 AEC Renown stands in the background.** R L Wilson

The acquired fleet was not only varied. Some of it was old; even very old. The oldest bus was nudging 23 years in service and some fleets had pursued uneven vehicle replacement programmes. If a 16-year operational life was the industry norm in the late 1960s (and many fleets aimed for less than that), Selnec had acquired over 200 time-expired buses from its constituents, representing a little under 10 per cent of its fleet.

The Southern fleet was the most modern, with an average age of 8.1 years, broadly equivalent to a 15 year operating life. The Central and Northern fleet ages averaged nine years, which represented a 17 year vehicle life. Selnec planned for a 13 year life, based on a major overhaul to get recertification for six years after the expiry of a bus's original seven year Certificate of Fitness.

Not all of Selnec's constituents had embraced the new technology of rear-engined buses with enthusiasm. Manchester had, and had been in the forefront of double-deck one-man operation. But it also suffered higher lost mileage than most other operators in the area, due in part to problems experienced in maintaining its complex rear-engined buses.

Some of the fleets – Leigh, Salford, Ramsbottom and Stockport – had been buying traditional half-cab front-engined buses until 1967 or later; indeed of these four only Salford had rear-engined vehicles. Ramsbottom with 12 buses could be excused this conservatism, but it was harder to excuse Stockport, with a fleet approaching the 150 mark. Indeed thanks to Ramsbottom, Selnec had the distinction of being the only PTE to take delivery of a new half-cab bus. The last home market Leyland Titan, ordered by Ramsbottom, was delivered to Selnec within days of its formation. But it was the Stockport inheritance which would ensure that the PTE was still running half-cab buses in the early 1980s.

The existence of the products of nine chassis makers was another area to be addressed. The chassis breakdown was:

Leyland	1,572
Daimler	787
AEC	118
Bedford	15
Bristol	6
Dennis	6
Atkinson	5
Guy	4
Albion	1

The bodywork variety was even more astounding and read like a who's who (or perhaps a who-was-who) of the coachbuilding industry – Alexander, Bond, Burlingham, Crossley, Duple Midland, East Lancs, Leyland, Longwell Green, Lydney, Marshall, Metro-Cammell, Neepsend, Northern Counties, Park Royal, Pennine, Plaxton, Roe, Weymann and Willowbrook. There were 19 in total although, as with the chassis, some of the more obscure were owned in only small numbers.

Add to this disparate collection of vehicles a wide range of operating practices, ticket issuing equipment, wage agreements, maintenance methods and scheduling techniques and some size of the mammoth task of integration which faced the new PTE gradually becomes apparent.

The livery was modern and the Selnec acronym was slick, although many opposed the change and the disappearance of long-established local fleets. But Selnec faced a major challenge in welding together its 11 constituents while still providing a reliable service for the 527 million passengers who travelled on its buses each year. Could the reality match the image?

Top left This remarkable veteran, complete with Selnec Central legal lettering, was a Park Royal-bodied AEC Regent which had been new to Salford in 1939. It had been transferred to the driver training fleet in 1949 and was still in stock 20 years later when Selnec took over. It is seen in Frederick Road depot in 1972 before being sold for preservation. R Marshall

Centre left Among the oldest buses to run in Selnec service were four 1950 Leyland Titan PD2s acquired from Ashton. They lasted into 1970 – this one was photographed in May – and received Selnec Southern legal lettering but were withdrawn without carrying their allotted fleet numbers, 5407-10. R L Wilson

Left By a strange quirk of fate one of Selnec's first new buses was a Leyland Titan, the last to be built for service in Britain. It had a 73-seat East Lancs body and had been ordered by Ramsbottom. It entered service in November 1969 in Ramsbottom livery and with fleet number 11. In this 1971 view it still carries the Ramsbottom crest but has been renumbered in the Selnec series. R L Wilson

Above Salford had bought a large fleet of Daimler CVG6s in 1950-51 and 48 of these were taken into Selnec stock. 4044, new in 1950, is seen in March 1970 soon after being renumbered. The 54-seat body is by Metro-Cammell and still carries Salford City Transport fleetnames. The short radiator, designed to minimise accident damage, was a feature peculiar to Salford and contributed to the particularly old-fashioned look of these buses. All were withdrawn in 1970-71; none were painted in Selnec livery. R L Wilson

SELNEC: THE INITIAL FLEET DIVISION

Model	Central	Northern	Southern	Total
Double-deck				
AEC Regent III	–	9	–	9
AEC Regent V	–	61	–	61
AEC Renown	–	18	–	18
Atkinson	–	–	1	1
Daimler CVG	368	66	26	460
Daimler Fleetline	230	49	35	314
Dennis Loline	–	6	–	6
Guy Arab IV	–	–	4	4
Leyland Titan PD1	–	–	1	1
Leyland Titan PD2	501	84	271	856
Leyland Titan PD3	–	77	36	113
Leyland Atlantean	304	124	69	497
Double-deck totals	1403	494	443	2340
Single-deck				
AEC Reliance	10	16	–	26
AEC Swift	–	4	–	4
Albion Nimbus	–	1	–	1
Atkinson	–	–	4	4
Bedford VAL	14	–	–	14
Bedford J2	–	1	–	1
Bristol RE	–	–	6	6
Daimler Fleetline	–	13	–	13
Leyland Royal Tiger	–	2	–	2
Leyland Tiger Cub	15	2	10	27
Leyland Leopard	–	10	5	15
Leyland Panther Cub	20	–	6	26
Leyland Panther	29	–	6	35
Single-deck totals	88	49	37	174
FLEET TOTALS	1491	543	480	**2514**

1970: Selnec Sets Sail

The new year dawned with the launch of a new service which epitomised integrated operation: the Trans-Lancs Express. This started in January and ran from Stockport to Bolton, linking Denton, Ashton, Oldham, Royton, Rochdale, Heywood and Bury on the way. It was launched by Southern using six 1968 Bedford VAL70 coaches transferred from the Manchester fleet and operated hourly with a through journey time of 1 hour 44 minutes and a fare of 6s (30p) for travel from one end to the other. The coaches had two-way radios.

The new Selnec orange and white livery was officially unveiled on 11th March, on 1968 Mancunian 1014. Selnec recognised the importance of promoting a modern image and 1014, which was less than two years old, was repainted specially for the launch. Routine repaints by this time which had received Selnec livery – all from the Manchester fleet – included 1965 Atlanteans 3789 and 3802; 1965 Fleetlines 4712/15 and 4720/27; 1967 Panther 53 and two PD2 Titans, 3615 and 3673. Six VAL coaches (207-212) were repainted in the new livery for use on the Trans-Lancs Express service. The oldest bus to feature among early repaints was ex-Manchester CVG6 4446 of 1954.

At the launch, no secret was made of the need to standardise. The target was to develop a standard Selnec bus by 1972.

Repainting actually started just in advance of fleet renumbering. Two buses, Rochdale Fleetline 328 and Oldham Atlantean 171, ran in Selnec livery with their old municipal fleet numbers before being given their new numbers, 6228 and 5171. Other early repaints into orange and white from Selnec's smaller constituents included Salford Daimler CVG6s 4024/28; SHMD Fleetline 5631 and CVG6 5682; Bury single-deck Fleetline 6089; Leigh lowbridge Leyland Titan PD2 6957 and Bolton Leyland Royal Tiger 6009.

Selnec fleet numbers were applied to vehicles with varying degrees of urgency. Most fleets did so using their existing style of fleet number transfers, as in Manchester, Bolton, Bury, Ramsbottom, Stockport and Rochdale. Manchester-style gold numbers were used on the ex-Salford fleet and appeared on former Ashton, Oldham and SHMD buses in the Selnec Southern fleet. This was because there were large stocks of Manchester number transfers to be used up. Most fleets were renumbered by the end of the year, although some Bolton buses lingered into 1971 without receiving their Selnec numbers, and only two Ashton buses – 5436/7 – were renumbered while still in blue.

In Leigh, the ornate shaded numerals which had been used to number the fleet would not have looked out of place on an Edwardian tramcar and renumbering was done using Manchester transfers. But when Leigh (and other Northern division) buses were repainted into Selnec livery, Bolton transfers were used to number vehicles. Ultimately the standard lettering for fleet number transfers was black Helvetica Light and this was eventually applied across the fleet.

The divisional structure was being put in place. The biggest, Central, was sub-divided into three districts. Two were in Manchester – the northern district based on Queens Road garage and the southern on Princess Road. The third, the north-western district, covered the former Salford City Transport operations and was controlled from Salford's Frederick Road garage.

Central had almost 1,500 buses, divided between seven former Manchester and two former Salford garages. The allocation at the start of 1970 was:

Manchester	
Birchfields Road	126
Hyde Road	236
Northenden	204
Parrs Wood	104
Princess Road	200
Queens Road	240
Rochdale Road	110
Salford	
Frederick Road	155
Weaste	115

Opposite **The Trans-Lancs Express was started in January 1970 and was run by Selnec Southern with stylish Plaxton-bodied Bedford VAL coaches. 211, one of six VALs delivered to Manchester in 1968, arrives in Stockport on its circuitous trip from Bolton in May 1970. Note the short-lived Selnec Express fleetname and the Trans-Lancs Express slip boards.**
R L Wilson

Left **The oldest ex-Manchester bus to receive Selnec livery was 4446, a 1954 Daimler CVG6 with Metro-Cammell body. It was repainted in 1970 and ran until 1973. It carries Selnec's imaginative private hire advertising.** R Marshall

Only two buses received Selnec livery and carried their old municipal fleet numbers. One was this ex-Oldham Atlantean, a 1968 Roe-bodied bus. John Aldridge

Different parts of Selnec applied new fleet numbers with differing degrees of urgency. In July 1970 Selnec Northern's 6638 was running with its Bolton fleet number, 138, neatly painted out – and then re-applied on a small handwritten sticker below the windscreen. It is one of ten Leyland Titan PD2s which entered service in 1961 and were the first Bolton buses with full-width cabs. Metro-Cammell built the 62-seat forward-entrance body. These buses were taken out of service in 1972-73. G R Mills

Each of the former municipal fleets in the Northern and Southern divisions were operated as districts with SHMD being described as Stalybridge, while Bury and Ramsbottom formed a single district.

The PTE started off with a forecast revenue deficit of over £2.5 million for the year 1970-71, created partly by the quirks of municipal accounting procedures and in particular their handling of depreciation and the funding of fleet renewals.

To address this the PTE proposed a 15 per cent fares rise, introduced in the Northern and Southern divisions on 8th February. Central's fares were to rise in two stages – in February and May. The February increase went ahead as planned, but delays in the delivery of new Sabloc fareboxes to further extend one-man-operation delayed the second rise, which introduced differential peak fares, until 3rd August. The differential fare structure – in effect a peak period surcharge – was not well received by passengers and was discontinued at the end of the year, to be replaced by a limited weekday cheap fare designed to encourage off-peak travel.

Unusually, Manchester had different fare scales for crew-operated and one-man services. On an omo bus 6d (2.5p) covered up to four stages (and was not increased in February) but the same journey on a crew-operated bus could cost up to 9d (4p) – which was increased in February to 1s (5p). Not only were omo fares generally lower, there were fewer of them to simplify change-giving. After the February increase the fares on omo buses rose in 6d (2.5p) increments from 6d to 2s (2.5p to 10p) while on crew buses fares rose in 3d (1p) increments from 6d to 2s 3d (2.5p to 11p).

In an effort to eliminate anomalies in the concessionary fares schemes inherited from Selnec's constituents, a new concessionary scheme was introduced in April. This gave local authorities the choice of buying half-fare passes for use on Selnec buses, or tokens which could be used on any bus in the area. This scheme was not entirely successful and was revised in January 1971.

At the start of the year revisions to the drivers' hours regulations forced new crew schedules on operators nationwide – for Selnec this proved to be a particular problem in Salford where duties had to be extensively reworked and additional staff recruited to maintain services under the new tighter drivers' hours rules. The changes saw Selnec lose a small amount of its operating mileage – under two per cent.

In April Selnec launched the Hale Barns

Above Selnec Central 3100 was one of 50 conservative Metro-Cammell-bodied vacuum-braked PD2/40s delivered to Salford in 1966-67 after the undertaking had gained experience running Atlanteans and Fleetlines. The bonnet is painted orange on the basis that orange is less likely than white to show-up any greasy finger prints left by fitters who have been working in the engine bay. The gantry of levers and handles above the radiator was fitted so that the conductor could change the destination and route number without climbing up on the bus, and was a standard Salford feature. R L Wilson

Left Ex-Bolton Atlantean 6685 of 1963 had just been recertified in the summer of 1970 and was among the early recipients of Selnec livery. The illuminated exterior advertisement was a short-lived 1960s feature. The styling of the East Lancs body on this bus was developed by Bolton general manager Ralph Bennett who then moved on to Manchester where he was responsible for the Mancunian body. The Bolton paintshop respected the body trim lines and gave it a shallower orange skirt than the Selnec standard livery specification called for. G R Mills

Express, an up-market commuter run from Hale Barns in Cheshire to Manchester city centre. This used the two newest Bedford VAL coaches in the fleet, repainted from Manchester's blue and white coach colours into Selnec livery.

A joint policy statement was approved by the PTA in June which set out a plan for future relationships with other operators. Under Section 20 of the 1968 Transport Act the PTA and PTE would approve the services and fares of British Rail in the area, and fund any revenue shortfall which resulted. Similarly the PTA and PTE planned to enter into an agreement with the National Bus Company so that NBC subsidiaries' routes could be integrated with the PTE's own bus operation. The PTE also had powers to run taxis, airports and garages, although these were not exercised.

No time was lost in starting consultation with other operators and by mid-year talks were in progress with British Rail, independent Lancashire United, and NBC subsidiaries North Western and Ribble.

These other operators accounted for less than 20 per cent of the 633.4 million trips made on public transport in the PTE area:

Passenger trips (millions)
Selnec PTE   526.9
National Bus Company   49.6
Lancashire United   34.4
British Rail   22.5

Negotiations had also been ongoing with the trades unions, leading to agreement in June to introduce uniform pay scales throughout the PTE for platform and maintenance staff. However within the uniform scales there were still numerous local differences – customs and practices built up over a long period of years could not be wiped out overnight.

New vehicles arrived in large numbers – all orders inherited from Selnec's predecessors. A total of 84 Leyland Atlanteans (the 11xx series) and 90 Daimler Fleetlines (21xx) ordered by Manchester had entered service by the end of 1970. All had stylish Mancunian bodies, built by Metro-Cammell, Park Royal and East Lancs. The East Lancs bodies, all on Atlanteans, included a batch of 12 single-doorway buses – the only Mancunians of this layout.

Atlanteans from 1161 and Fleetlines from 2151 were delivered in Selnec livery with a new specification interior trim and had forward-ascending staircases, a feature also specified on 1131-54, the East Lancs-bodied buses. This layout was considered to be safer than the rearward-ascending staircases of the original Mancunians.

Above **The Hale Barns Express was launched in April 1970 and used two 1969 Bedford VALs with Plaxton Panorama Elite 52-seat bodywork. The coaches were fitted with tables. 214 picks up outside the Tatler Cinema Club in Whitworth Street West where tired executives could stop and view Chuck Connors in a Ride Beyond Vengeance as an alternative to a ride home to Hale Barns.**
R Marshall

Right **The flow of new Mancunians which had been ordered by Manchester continued in 1970 and included the last of the Atlanteans; deliveries of Fleetline-based Mancunians continued until 1972. 1118 was a Park Royal-bodied PDR2 with 75 seats and room for 25 standees, making it a genuine 100-passenger bus – in theory at any rate. Whatever happened to Fennings Little Healers?**
R L Wilson

A Salford order for 20 Metro-Cammell bodied Atlanteans was built to Mancunian specification but with Salford's style of destination display and green Salford seat coverings. These buses, delivered in Selnec livery as 1201-20, also had Salford registrations, SRJ324-43H, which matched their intended Salford fleet numbers.

Smaller batches of new buses with obvious municipal origins were delivered to Southern and Northern. Southern received five Roe-bodied Atlanteans to full Oldham specification early in 1970. These entered service in Oldham livery and like the Salford Atlanteans had local registrations, WBU183-7H, which matched their intended municipal fleet numbers. They were numbered 5183-87 by Selnec.

An outstanding Ashton order for four Northern Counties-bodied Atlanteans was delivered in Selnec livery but the vehicles were otherwise to Ashton specification. Here again the registration numbers, VTE162-5H, matched the intended fleet numbers, 62-65. In allocating numbers to the Ashton fleet Selnec had left 5462-65 vacant in anticipation of the delivery of these buses.

Ten new Bristol VRTs had been ordered by Stockport for delivery in 1970 and would have been that fleet's first rear-engined buses – but they were destroyed by a fire at the East Lancs factory in Blackburn in April. They would have been numbered 5898-5907 by Selnec. Consideration had been given to allocating the VRTs to Leigh for use as omo buses. The garage at Leigh was unable to accommodate full-height 'deckers but could have housed the lowheight Bristols. Some of the VRT chassis survived the fire but Selnec did not want them and they were exported to Australia.

The Northern division received seven East Lancs-bodied Fleetlines ordered by Bury. These were delivered in Selnec livery as 6344-50 but were built to Bury's specification. Once again the registration numbers, NEN 504-10J, gave the clue to their intended municipal fleet numbers.

Bolton was in the process of receiving 15 two-door East Lancs-bodied Atlanteans as the undertaking was being transferred to Selnec. Only four of these, Bolton's 287, 291, 295 and 300, had been delivered by 1st November 1969. The remaining 11 were delivered to Selnec but were of course to full Bolton specification and in Bolton livery, the only concession to the new regime being the absence of municipal crests.

East Lancs was also bodying Manchester Atlanteans 1131-54 when the PTE was starting up. Six (1131/3-5/8/41) were delivered to Manchester City Transport; the balance to Selnec. Another vehicle in build at East Lancs as Selnec was taking over was Ramsbottom's 11, the last Leyland Titan built for operation in the United Kingdom. It arrived at Ramsbottom shortly after the takeover.

Rochdale was committed to buying a demonstration Fleetline SRG6LX single-decker from Daimler and this bus, KKV700G, was delivered to Northern in March 1970 in Rochdale livery and numbered 6038. An order placed by Rochdale for ten similar chassis with Strachan bodywork was cancelled by Selnec.

The influx of new buses was matched by withdrawals of old ones. By the end of the year the vast majority of Manchester's oldest Titans had gone, JND- and NNB-registered buses new between 1950 and 1953, although a few 1951 PD2s survived into 1971. None of these buses, which were numbered in the series 3200 to 3364, were repainted into Selnec livery. All of the Daimlers delivered during 1950-51 to Manchester were withdrawn in 1969-70, along with most of Salford's Daimlers of the same era; a few of Salford's early Daimlers survived to see service in 1971. None received Selnec livery. At the end of the year a few ex-Manchester TNA-registered PD2s in the 34xx-series received partial red repaints and at the same time Selnec logos and Central fleetnames.

Opposite top **The Metro-Cammell bodies on an outstanding Salford order for 20 long-wheelbase PDR2 Atlanteans were to Mancunian specification, but were immediately identifiable by their Salford-style destination screens. They entered service in 1970. The Mancunian body had translucent panels for most of the length of the roof, which brightened up the interior quite considerably.** R L Wilson

Opposite centre **Selnec Southern's first new buses were six Roe-bodied Atlanteans which had been ordered by Oldham and were delivered in Oldham livery. Oldham had standardised on Atlanteans for its double-deck fleet since 1965 and, like many other operators, had specified dual-door bodies from 1969 when double-deck one-man-operation was legalised.** R L Wilson

Opposite bottom **By the time an outstanding Ashton order for four Atlanteans was delivered the Selnec livery had been chosen and the new Ashton-style buses arrived from Northern Counties in orange and white. 5462 is seen standing in for a coach on the Trans-Lancs Express in May 1970.** R L Wilson

This page top **A Bury order for seven East Lancs-bodied Fleetlines was delivered to Selnec Northern in 1970. The first of these, 6344, is running as a one-man bus in Bury town centre.** R L Wilson

Above **Rochdale had four single-deck Fleetline SRG6LXs with Willowbrook bodies. 6030 is seen amid Regents in Rochdale centre.** R L Wilson

Ageing Titans at Oldham were also quick to go and all 19 pre-1952 buses were out of service (again without being repainted) by the end of 1971. At the same time a start was made on withdrawing the 86 Titans which survived from deliveries made to Oldham between 1954 and 1958, although withdrawal of these more modern buses was spread over the following six years. Metro-Cammell bodied buses were generally singled out for earlier withdrawal than those bodied by Roe. Out of 15 Metro-Cammell bodied Titans only three survived to receive Selnec livery.

From the Ashton fleet 1950 Titans 5407-10 were quickly withdrawn – which, as they were 20 years old, was hardly a surprise. But Ashton's next-newest Titans, 1955 buses 5411-17, had also gone by the end of 1970. None were repainted out of their old colours. Bury's two non-standard AEC Regent IIIs, 6376/7, which were also the oldest buses acquired with that fleet, were withdrawn in 1970, still in green.

Opposite top **One of the early Oldham repaints was Titan 5329, a 1958 Roe-bodied PD2. It was part of a batch of 24 similar buses and was withdrawn in 1975. Although the bus is in full Selnec Southern livery the destination display serves as a reminder of the old regime: it reads 'Oldham Transport'.**
G R Mills

Opposite bottom **Rochdale's standard double-decker in the 1960s was the Daimler Fleetline. Three batches were purchased, totalling 22 in all. The 1965 delivery had 77-seat Weymann bodies. 6229 was repainted in Selnec livery in the spring of 1970 and served the PTE until 1979.** R L Wilson

Top left **New for the Trans-Lancs Express was this Seddon Pennine IV, one of eight delivered to Selnec in the spring of 1970. The Pennine IV had a front-mounted Perkins engine and Selnec opted for Plaxton Panorama Elite bodywork. This is Bolton's Moor Lane bus station. The 1970 Seddons were short-lived and were sold in 1973.**
G R Mills

Centre left **To introduce one-man operation to Stockport, 13 ex-Manchester Atlanteans were transferred from Selnec Central to Selnec Southern in 1970. These were 1966 PDR1/2s with 75-seat Metro-Cammell bodies. 3822 carries the Southern flash and an advert for what was then a little-known make of Japanese car. The PDR1/2 had a drop-centre rear axle and an SCG gearbox.**
R Marshall

Bottom left **The strangest coach in the Selnec fleet was 215, the ex-Salford AEC Reliance with Weymann's attractive Fanfare body. It was put to work on the service to the Airport, repainted turquoise and white but devoid of Selnec fleetnames.** R L Wilson

With an eye to expanding coach operations nine new vehicles joined the coach fleet in 1970. Eight were Seddon Pennine IVs (217-224) with 170bhp Perkins V8.510 8.36-litre engines which supported local industry (the Seddon works were in Oldham) but added yet further variety to an already varied fleet. The ninth was an odd machine too, a Bedford VAS5 (216). All had Plaxton bodies. Six of the Seddons were allocated to Ashton for use on the Trans-Lancs Express. The other two went to Parrs Wood.

With the arrival of the Seddons four of the displaced Bedford VALs were allocated to the Northern division for use on private hires and as spares for the Trans-Lancs Express. Two were kept at Leigh and one each at Bolton and Bury.

The 14-strong Bedford coach fleet inherited from Manchester City Transport had one other vehicle added to it. This was Salford's municipal committee coach, a 1962 AEC Reliance with luxurious Weymann Fanfare bodywork with two-and-one seating for only 26 passengers. It was painted into the former Manchester coach livery of turquoise and white, transferred to Parrs Wood and used on service 200 to Manchester Airport. It was withdrawn in 1971.

Inter-division transfers soon became a feature of Selnec, although by and large few buses strayed far from their town of origin. To compensate for the loss of the Bristol VRTs in the East Lancs fire a batch of 13 1966 Atlanteans, 3810-22, was transferred to Stockport (Southern division) from Manchester (Central division). They introduced double-deck one-man-operation to the town. Three of them also ran briefly at Ashton, another Southern division depot.

Some of Stockport's 1951 PD2s featured in transfers within the Southern division. Two, 297/8, were moved to Stalybridge where they received SHMD fleet numbers 51 and 52. This was in February, prior to the introduction of the Selnec fleet numbering system and livery, and 52 was actually repainted in SHMD green. Eight other 1951 Stockport PD2s went to Stalybridge but retained their red livery and Stockport fleet numbers. The 1951 PD2s replaced the only Daimler-engined buses in the Selnec fleet, six 1952 ex-SHMD CVD6s, along with a couple of Atkinsons and a fire-damaged Fleetline.

Stockport PD2 302 went to Oldham where it received Oldham livery and was then numbered incorrectly as 5202 in the series used for ex-Oldham buses before becoming 5922 in the ex-Stockport series. The number 5202 did have a certain logic about it: to bring them into the Selnec series 4900 was being added to the fleet numbers of ex-Oldham buses numbered above 300.

Top **Stockport 297, which became Selnec 5917, made its way to Stalybridge in March 1970 where it ran with fleet number 51 in the SHMD series. It is seen here in July 1970 in the company of one of only four Guys to be acquired by Selnec, ex-Ashton Arab IV 68. The Arabs were new in 1956 and had Bond bodies. They were withdrawn later in 1970 without being repainted or carrying their new fleet numbers.** G R Mills

Above **Stockport 298 also moved to Stalybridge and assumed the guise of SHMD 52, right down to the green and cream SHMD livery. This bus dated from 1951 and had little life left in it. It is hard not to interpret the green repaint as someone's last ditch defiance of a new authority.** N R Knight

Left **The third Stockport bus to assume a strange identity was 302 which was transferred to Oldham and repainted in Oldham livery. It ran briefly as 5202 in the ex-Oldham number series before receiving its correct number, 5922.** R Marshall

More modern Stockport Titans, PD2As 5801-10 new in 1963, were also moved around as a result of the omo conversions introduced with the ex-Manchester Atlanteans. Oldham received six while four went to Ashton. These buses featured frequently in transfers in the early 1970s with all ten running in Stalybridge during 1971, before some gravitated to Manchester garages in the middle of the decade. Oldham Atlanteans 5127 and 5138 went to Stockport for use as omo buses.

A review of Central's vehicle requirements led to a re-assessment of the need for single-deckers and the 15 ex-Manchester Leyland Tiger Cubs – three of which were in blue dual-purpose livery when Selnec took over – were withdrawn by Central in 1970. The nine single-doorway vehicles were sold but the six dual-door buses, 10-15 in Manchester red livery, were moved to Oldham where they operated until 1975, latterly with Southern fleet numbers 5000-5. Oldham already had a fleet of 17 two-door single-deckers in operation, including six Tiger Cubs.

At the same time all but two of the 20 Panther Cubs acquired from Manchester were withdrawn – many finding new owners in Australia. The odd two, 21/22, were repainted in Selnec livery (as had been 29) – but only survived until March 1971. They ran briefly in Salford, along with a number of ex-Manchester Panthers which ousted the nine ex-Salford AEC Reliance buses, 65-73, which were only eight years old.

Top **Transfers within Selnec Southern in 1970, caused by the introduction of one-man-operated double-deckers to Stockport, included a batch of 1963 Stockport PD2A Titans which were shared between Oldham and Ashton. 5810 is seen in Yorkshire Street, Oldham with Oldham Corporation route number and destination blinds.** R Marshall

Centre **There were 15 Leyland Tiger Cubs in the Manchester fleet when Selnec was formed. Nine single-door buses were withdrawn in 1970, but six 1962 PSUC1/2s with 38-seat dual-door Park Royal bodies lasted rather longer, despite having constant-mesh gearboxes which can hardly have endeared them to drivers on urban routes. They were transferred to Oldham in 1970 and operated there until 1975. In this 1972 view ex-Manchester 59 still carries its Selnec Central number, 14. It was renumbered 5004 in the Selnec Southern series in 1973. The Oldham Corporation destination screen was a poor fit.** R L Wilson

Left **The Panther Cubs acquired from Manchester were quickly withdrawn by Selnec – but not before three had been repainted, including 22, seen here in Weaste on an ex-Salford City Transport service in January 1971. It was withdrawn later that year. New in 1965 it had a 43-seat Park Royal body.** R L Wilson

Salford's nine AEC Reliance buses were short-lived and were withdrawn in 1970-71 after less than ten years in service. As with Salford's oldest Daimlers, none were repainted – but unlike the ageing Daimlers the Reliances soon found new homes with other bus operators. Weymann built the bodywork. A dirty 66 leaves Victoria bus station for Peel Green in April 1970. *R L Wilson*

Selnec's only Albion was a stylish Weymann-bodied Nimbus which came with the Ramsbottom fleet but had been new in 1963 to Halifax and had also run for a period with Warrington. This neat little bus was a 31-seater and ran until 1976. It is seen leaving Ramsbottom depot. *R L Wilson*

In the Northern division the sole Albion in the Selnec fleet, a one time Halifax Nimbus operated by Ramsbottom, was repainted orange and white in 1970. While this was being done its duties were covered by former Bolton Royal Tiger 6009.

The only Guys in the Selnec fleet, ex-Ashton 5466-69, survived rather better than might have been expected and while 20-year-old ex-Ashton Titans were being withdrawn the Guys managed to see some all-day service before they too were taken out of use. Two 1957 PD2s from Oldham, 5288/9, were drafted in to Ashton to replace withdrawn buses.

Rochdale Road depot in Manchester closed on 24th August, following a review of Central division garage requirements. There was also concern that there was a risk of subsidence from long-disused colliery workings in the area. The services operated by Rochdale Road were covered by Hyde Road, which took over around two-thirds, and Queens Road which got the remainder.

A new Stockport garage and head office for the Southern division was formally opened at Daw Bank in October. This had been started by Stockport in municipal days and allowed the Heaton Lane garage acquired from Stockport Corporation to be closed.

The most significant event for the future of Selnec's vehicle policy was the appearance at the Commercial Motor Show in September of a prototype standard bus. This was a Northern Counties bodied Leyland Atlantean PDR1A/1, one of six ordered by Ashton in 1969. The vehicle was evolved by Selnec engineers working in close consultation with Northern Counties, who had not been a major body supplier to Manchester City Transport, although some of Selnec's smaller constituents had been customers of the Wigan-based company. A combination of competitive pricing and its close proximity to Manchester helped it win the prototype orders – and large production volumes thereafter.

The aim of Selnec's engineers was to build a vehicle which would be attractive both to passengers and drivers and would also offer maintenance advantages over existing types. The starting point for the design was the standard Park Royal group double-deck body which had 5ft 4in pillar spacing. This meant that the position of the main body side pillars no longer corresponded with the chassis outriggers, a fact which was to become a source of trouble on Northern Counties bodies in later years.

The use of a standard wheelbase PDR1 rather than the long-wheelbase PDR2 favoured by Manchester City Transport was designed to produce a lighter and more manoeuvrable bus than the MCT Mancunian. It had only a single entrance/exit door because the two-door layout used on the Mancunians had not produced any appreciable saving in boarding and alighting times at bus stops.

A forward-ascending staircase, as used on the later Mancunians, was fitted. A two-step entrance, colourfully described in the Selnec staff newspaper, Selnec Express, as a "two-step tramcar entrance" was used. The engine shrouds favoured by Manchester for the Mancunian were abandoned.

The driver's cab had the handbrake and gear selector mounted on the right to leave the area on the left free for fare collection equipment. This was truly innovative thinking – most buses at this time had the gear selector on the driver's left. The driver had a heated windscreen. Selnec did a lot of work on the electrical layout with plug-and-socket connections to ease fault finding and replacement.

The show bus, EX1, started a short-lived experimental series of numbers – more of which later. After the Show Selnec announced its first major bus order – 500 of the new standards were to be built by Northern Counties (300) and Park Royal (200) on Daimler Fleetline (350) and Leyland Atlantean (150) chassis. Selnec had intended to divide the body order equally, but Park Royal was unable to supply 250 bodies in the time that was available.

Although anxious to develop its own standard bus Selnec was also interested in developments elsewhere. A Metro-Scania demonstrator, VWD451H, was borrowed and used in service in the Central division from Princess Road garage in October.

This bus came under the scrutiny of Selnec's engineers who recorded that on service 53, the so-called circular from Queens Road garage to Old Trafford, it returned 4.79mpg compared with 6.46mpg for Atlanteans and 6.59mpg for Fleetlines on the same route. It was also tried on the 123 from Manchester to Greenheys where its fuel consumption of 4.78mpg compared unfavourably with 7.2mpg for the Leyland Panthers on the service. Based on then current fuel prices and rebates the Metro-Scania's annual fuel bill would have been £246 more than that of a Fleetline, assuming it covered 29,000 miles a year, the Central division's fleet average.

Part of the difference in fuel economy could be attributed to the lively performance of the Metro-Scania. When tested unladen by Selnec engineers it accelerated from 0-30 mph in a sprightly 13.2 seconds, compared with 17.2 seconds for a PDR2 Atlantean.

At the end of the year the Department of Trade and Industry announced that it was sponsoring the manufacture of two battery-electric buses which would be loaned to interested operators – one of which was Selnec, which was later to take a lead in developing battery-powered buses. The DTI expected the buses to be ready in the second half of 1971. In the event Selnec had to wait until 1973 to try one.

In October BR closed the Rochdale-Bury-Bolton line and extra peak journeys were added to the Trans-Lancs Express between Bolton and Rochdale. At the same time the PTE decided to seek parliamentary powers to improve the rail network by building an underground link in central Manchester which would connect the city's two main terminals at Piccadilly and Victoria. Selnec's rail plans also included a rapid transit line from Northenden to Blackley, reviving recommendations made to Manchester City Council by consultants in the mid-1960s, and the building of 66 bus/rail interchanges.

Fares rose again in December with Selnec being accused of having put fares up by 24 per cent since its formation.

During the 14 months to December 1970 Selnec's buses carried over 623 million people while rail services in the PTE area carried 23 million. The bus operation incurred a £179,200 deficit on receipts of over £23.5 million for the period, which was attributed to pruning of the January 1970 fares increase and delays in implementing the fares rise in the Central division. At the end of the year the fleet stood at 2,439 buses of which 1,125 were suitable for omo.

Conversion to one-man-operation, with the loss of conductors' jobs, was a thorny issue and although 46 per cent of the fleet was suitable for omo, only around 27 per cent of Selnec's operation actually was omo, mostly using double-deckers. One-man-operation was being extended to alleviate staff shortages as well as to reduce rapidly rising costs. At the end of the year, for example, there were conversions in Stockport and in Rochdale. The latter used ten former Rochdale Fleetlines (6235-44) fitted with fareboxes on service 21, Rochdale to Bury via Heywood. This was Rochdale's first double-deck omo.

At the end of the year Selnec employed 10,252 people of whom 4,013 were drivers and 2,553 were conductors. The route network covered 1,043.51 miles made up of 992.63 miles of local bus services and 50.88 miles of express services.

Red-liveried Metro-Scania demonstrator VWD451H was operated from Princess Road for four weeks.
K Walker

1971: Restructuring

The year 1971 started with an operating agreement being reached with Lancashire United Transport. LUT was Britain's biggest independent operator with a fleet of 375 buses and coaches. Its operating area stretched from Manchester to Liverpool but its base at Atherton and many of its services were in the Selnec area. The agreement allocated LUT a proportion of the bus mileage in the PTA area to be operated at an agreed rate and on which fares would be fixed by the PTE and PTA.

The 1970 concessionary fares scheme was replaced from 1st January by a half-fare elderly persons pass, valid throughout the Selnec area – but only for off-peak use. From 1st January 1972 the scheme was widened to allow half-fare travel at all times.

A significant restructuring took effect on 1st March with the three divisions becoming separate companies under the Selnec Bus Holding Company. The divisions became Selnec Central Bus Company, Selnec Southern Bus Company and Selnec Northern Bus Company. Each company was given its own director/general manager. The holding company was headed by Richard Cochrane, Selnec Central by Jack Thompson (formerly general manager at Manchester), Selnec Southern by Norman Kay (who had been general manager at Bury) and Selnec Northern by Jim Batty (formerly general manager at Bolton).

The original intention was to form the companies as Central Bus (Selnec) Ltd, Southern Bus (Selnec) Ltd and Northern Bus (Selnec) Ltd – until it was established that PTAs could not have subsidiaries which were limited liability companies.

The PTE's bus route network in 1971 covered just over 990 miles of road and the average length of the 460 basic services was 6.8 miles. Nine services were more than 15 miles long; 21 were under two miles in length. The average scheduled speed of Selnec's buses was in the region of 12 to 13mph. Average bus running costs were 35p a mile or £3.70 an hour.

In March the traffic management system in Manchester city centre was extended southwards from Piccadilly to Whitworth Street. To avoid disrupting inbound bus services two short stretches of bus lane were introduced in London Road, near Piccadilly Station – the first in the city.

Costs were still rising and in August the PTE applied to the Traffic Commissioners (who controlled fares through the Road Service Licensing system) for a 15 per cent fares increase, designed to cover inflation. This was reduced by about a third as the Commissioners took cognisance of the government's anti-inflationary policies, leaving the PTE with an expected deficit of £878,000 on its 1971 bus operations. However an increase in the government's new bus grant (from 25 to 50 per cent of the purchase price of an approved bus) coupled with operating economies, cut the deficit to £184,000.

New vehicles in 1971 were still largely the residue of outstanding orders from Selnec's predecessors. Additions to the coach fleet were similar to those in 1970 – a further eight Seddon Pennine IVs (227-234) and two more Bedford VAS (225/6), again all bodied by Plaxton. The 1971 Bedfords were small-engined VAS1 models (4.9-litre, 97bhp), unlike the 1970 coach which had been of the bigger-engined VAS5 variety (5.4-litre, 105bhp). These vehicles were shared between all three companies.

Two demonstration coaches were evaluated during the year. Bedford's BXE285J, a mid-engined YRQ with Duple Viceroy body, was operated by Central and Southern in the spring. It made a good enough impression for an order to be placed for eight for 1972 delivery. A Ford R-series, KEV950J, ran for the same two companies in August, but Ford was never to supply the Selnec coach fleet with anything bigger than a Transit.

The flow of Mancunians continued at a slower rate, with 84 Fleetlines being delivered. The first 14 were the balance of a 1970 delivery (2151-2210) with Metro-Cammell bodies; the remaining 60 (2211-70) had Park Royal bodies. All had been ordered by Manchester Corporation and were delivered to Selnec Central.

Selnec Southern received the last outstanding Oldham order, 12 Atlanteans with dual-doorway Roe bodies (5188-99). These were built largely to Oldham specification but without engine shrouds and were delivered in Selnec livery. They had Oldham registration numbers which matched the last three digits of their fleet numbers.

A small number of new single-deck buses arrived in 1971, and these too were a legacy from a now defunct municipal operator: Pennine-bodied AEC Swifts ordered by Rochdale. The order called for ten (6040-49) of which four were delivered to Selnec Northern by the end of the year. Again they had local registration numbers which matched their fleet numbers.

Opposite Five Leyland Leopards in the Bolton fleet went to Selnec Northern. 6054, a 36ft-long PSU3, pulls out of Bolton bus station in August 1971 soon after receiving Selnec livery. The 49-seat body is by East Lancs and is unusual in being of two-door layout on a vehicle built as early as 1964. The bus is heading for Belmont on the edge of the West Pennine Moors. It passed to GMT and was withdrawn in 1977. R Marshall

Left This unusual beast was an earlier generation of Bolton single-decker: a Bond-bodied Leyland Royal Tiger PSU1. It was new in 1956 and by the time it passed to Selnec had been equipped for one-man-operation. It received its Selnec fleet number, 6010, but was withdrawn in 1972 without being repainted. R L Wilson

Above left Bolton's Daimler CVG6s with Metro-Cammell Orion bodies were handsome buses which looked well in the new Selnec livery. There were 11 in all, most of which were withdrawn in 1973, although one survived until 1975. 6596 passes 6646, a 1960 CVG6-30 still in Bolton livery, in August 1971. R Marshall

Above right Selnec Northern was the only part of the PTE to run AEC Regent Vs, starting with a fleet of 61. Most were ex-Rochdale buses, but six came from Bolton and were the only 30ft-long Regents in the Selnec fleet. New in 1961, they had 72-seat forward-entrance Metro-Cammell bodies. R Marshall

Left Perhaps the most ungainly of Selnec's acquired Titans were 17 ex-Bolton PD3As with asymmetrical full-fronted bodies by East Lancs and Metro-Cammell. The deep nearside windscreen made full use of the cut-away in the PD3A's bonnet, but did little for the overall appearance of the bus. These buses were new in 1962-63 and were withdrawn in 1975-76. 6680 had a Metro-Cammell body. To meet the Selnec standard livery layout the orange around the upper deck windows should have been extended downwards a few inches, as on the Regent V in the previous picture. R Marshall

Another outstanding municipal order delivered in the winter of 1971-72 was Bolton's last batch of Atlanteans. These were long-wheelbase PDR2/1s with 86-seat East Lancs bodies. There were 15 in all – 6802-16, with Bolton registrations – of which six were delivered by the end of 1971. The bodies were to a fairly typical East Lancs design but with Selnec interior trim and destination screen layout. 6802-12 went to Bolton but the last four, 6813-16, were allocated to Bury.

EX1, the 1970 Show exhibit and prototype Selnec standard double-decker, entered service with Central at Queens Road in March. It was soon followed by identical EX2-6. All had 75-seat single-doorway Northern Counties bodies on Leyland Atlantean PDR1A/1 chassis which had been ordered by Ashton. Two were allocated to each of the companies – EX1/2, Central (at Queens Road), EX3/4, Southern and EX5/6, Northern. All had Selnec fleetnames with an orange flash, rather than divisional identities. To compensate Southern for the loss of its new buses, Central loaned Mancunians 1132/33 to Ashton, while Northern sent Bury-style Fleetlines 6346/47. EX1-6 were finally brought together at Ashton early in 1972 and 1132/33 and 6346/47 were returned.

While these new buses were being delivered Selnec's engineers were looking at ways of reducing maintenance problems on rear-engined buses. A new inspection system had been introduced in 1970 and this, along with increased expenditure on maintenance, had improved vehicle availability and cut lost mileage. But as a long-term development Selnec was examining the concept of a removable power-pack which if it became defective could quickly be detached from the body and replaced by a spare. A patent was taken out on this idea. And, in its frustration, there was even a hope that British Leyland could be talked into restarting Guy Arab production if a big enough order was promised. But neither the forward-looking power-pack nor the backward-looking Arab came to pass. British Leyland's response was that the Arab jigs had been scrapped.

The first Selnec order for new single-deck buses was announced in the summer – but called for only 12 vehicles, indicative of the lack of real interest in single-deckers in an area where double-deckers had always been dominant. These were Metro-Scanias, and they were ordered following the month-long demonstration of prototype VWD451H in the autumn of 1970. Eight were the standard 11m-long model and four were to be 10m long, the first order placed for short versions. The 11m models were delivered in 1972 as EX42-49 while the 10m buses did not arrive until 1973 and were numbered EX50-53.

Transfers brought odd buses to Ashton in the shape of three of Oldham's new Roe-bodied Atlanteans, 5190-92, along with similar 1969 buses 5181/2. Two of Bury's ageing Titans, 6380/6 of 1953, were also transferred from Selnec Northern to Selnec Southern at Stalybridge where they were withdrawn in 1972, still in Bury livery.

Above Rochdale had made small-scale use of omo single-deckers for some years. 6021 was a 1964 AEC Reliance with 43-seat East Lancs bodywork, one of three. It is flanked in the town centre by Swift 6037 and Fleetline 6032. The Reliance was withdrawn in 1976. *R Marshall*

Below A lot of thought went into the design of a Selnec standard body and the first six prototypes, EX1-6, were built in the winter of 1970-71 by Northern Counties on PDR1A/1 Atlantean chassis which had been ordered by Ashton. The result was crisp and practical – albeit less visually exciting than the Mancunian. The first six of the 21 EX-series prototype double-deckers had the route number display in the offside box on the front, which had been Manchester practice. Later EXs and all production standards had the route number on the nearside where it could more easily be seen from the kerb if a line of buses was approaching a stop. *P Sykes*

SHMD switched from Daimlers to Leylands in 1958 and its first-ever examples of the make were eight PD2s with Northern Counties bodies. Eight more followed in 1959 and 1962. 5689 was one of the first and operated for Selnec until 1972. The last of the ex-SHMD Titans were withdrawn in 1975. R L Wilson

1963 YDB-registered PD2A Titans 5801-10 which had been transferred from Stockport to Ashton and Oldham in 1970 moved on to Stalybridge in 1971. Other ex-Stockport PD2s on the move within the Southern company included 1958 NDB-registered Crossley-bodied 5938/39 which spent time at Ashton, and 5940/41 which had a spell at Stalybridge along with 1967 PD2s 5851-53. Four of the five G-registered rear-entrance Stockport PD3s were moved to Oldham, along with 1958 buses 5936/7.

The oldest surviving ex-Stockport Titans, ten 1951 all-Leyland PD2s, were taken out of service in 1971 along with the oldest ex-Leigh buses: 1950 PD2s with lowbridge Lydney bodywork. None of these venerable machines received Selnec livery. To speed the withdrawal of Salford's 1951 Daimlers, a dozen of which survived into 1971, former Manchester CVG6s 4409-20 were drafted into Salford garages as replacements. These NNB-registered buses were in fact only slightly younger than the ex-Salford vehicles which they replaced, although with their new-look fronts they looked much more modern than the exposed radiator Salford buses. Later in the year more Manchester buses made their way to Salford when five 38xx Atlanteans and five 47xx Fleetlines were allocated to Frederick Road and Weaste respectively as part of an omo conversion programme. Transfers within the Central company between ex-Manchester and ex-Salford garages became quite common.

The conversion to omo of route 9 from Ashton to Rochdale via Oldham saw Fleetlines on the move with two ex-Bury Alexander-bodied buses going to Rochdale and three ex-SHMD Northern Counties-bodied examples being allocated to Ashton. Former Ashton PD2s 5418-21 of 1960 were then transferred to Stalybridge – transfers involving the Ashton fleet were rare.

Repainting of the fleet in Selnec livery was continuing while many of the older buses were withdrawn without being repainted. The only exception to the new livery rule in 1971 occurred at Stalybridge where at the end of the year eight ex-SHMD Leyland Titans, 5601/3/5 of 1962 and 5685/88-90/92 of 1958, were given partial green repaints to use stocks of SHMD green paint. This laudable attempt to minimise costs by making full use of available materials must have been met with mixed feelings at Selnec headquarters.

Late summer saw Selnec hiring a fleet of 11 new Seddon Pennine RUs from Lancashire United Transport for evaluation. LUT's 382-392 (DTC730-740J) were divided between the three companies. Southern used 382/3 at Stockport while Northern had 384-7 which were initially operated by Bolton, Ramsbottom, Leigh and Rochdale although all four ended up at Bolton. The last five were operated by Selnec Central from Queens Road, primarily on the 147 from Cannon Street to Hollinwood. All had two-door Plaxton bus bodies and ran for Selnec during July and August.

Their appearance on the 147 followed an incident earlier in the year when frustration with the notoriously unreliable Panthers used on the route saw it being converted to double-deck operation – which meant turning 300 yards short of the Hollinwood terminus which lay beyond a low railway bridge. This caused some complaint and the Panthers were soon re-instated.

Manchester's Parrs Wood garage was closed to buses in April – but was later to have a new lease of life as a depot for Selnec Parcels. Most of the Parrs Wood operations were relocated to Birchfields Road.

Left Stockport's PD3s did not figure strongly in transfers but 5891 did find itself working from Oldham for a spell in the early 1970s. New in 1969, it was the last rear-entrance bus purchased by Stockport. S Burton collection

Right Manchester had a large fleet of relatively young Leyland Titan PD2s which were to serve the city under PTE ownership until the mid-1970s. This PD2/40 with Metro-Cammell 65-seat body was one of 100 similar buses delivered in 1958-59. Just over half of this batch of buses was withdrawn in 1971-72, but the remainder survived until later in the decade. This bus, 3540, was withdrawn in 1976. G R Mills

Much was happening on the rail front. In July it was announced that under Section 20 of the 1968 Transport Act the PTA and PTE would be assuming responsibility for reviewing and integrating rail services in the area as part of the total transport system. The review – of 25 services on 19 lines – was soon under way and concluded that the system needed considerable capital expenditure to modernise stations and replace time-expired rolling stock. Time-expired some rolling stock may have been, but 14 years were to pass before any new trains actually arrived on Manchester's local rail lines. Financial responsibility for the rail services was to pass to the PTA and PTE from 1st January 1972.

As this was happening preparatory work was under way on the Manchester Central Area Railway Tunnel, designed to link the rail networks to the north and south of the city and soon to be known as the Picc-Vic line. An enabling Bill for the link was lodged with Parliament in November.

The underground line was to have five stations – Piccadilly Low Level, Princess Street/Whitworth Street, St Peters Square/Albert Square, Market Street and Victoria Low Level – and would cost £26.9 million. PTE director general Tony Harrison saw it as part of a grand integrated transport scheme for Greater Manchester with bus priorities and bus/rail feeder services.

The electrified Altrincham to Manchester line was converted from 1500V dc to 25kV ac from May and services were linked with the Manchester to Crewe line running via Deansgate, Oxford Road and Piccadilly stations. The Altrincham line had originally been electrified by the Manchester South Junction & Altrincham Railway in 1931 and was the first in the country to use the 1500V dc overhead system. The original Manchester South Junction & Altrincham-style electric multiple units operated on the line until its conversion to 25kV when they were replaced by the AM4 (later known as Class 304) units already in use on the Crewe line.

Top Most of Oldham's ten 1958 Metro-Cammell-bodied PD2s were withdrawn in 1971 without being repainted: 5322 is seen near the end of its days, looking remarkably respectable for a bus which was soon to be consigned to the scrap heap. The non-standard grille was an Oldham peculiarity. R L Wilson

Centre Another type which did not receive Selnec livery was the AEC Regent III. Leigh provided seven, all new in 1952 and all with 53-seat lowbridge East Lancs bodywork. 6940 was withdrawn in 1971 – this photograph was taken in May of that year – but some of the batch lasted until 1973. R Marshall

Right Leigh contributed the only AEC Renowns to be found in the original Selnec fleet although more were to appear in 1972 with the North Western takeover. This 1964 Renown with 72-seat East Lancs body is representative of 14 delivered to Leigh between 1964 and 1967. Two earlier examples had rear-entrance bodies. All of the Leigh Renowns passed to GMT and the last survived until 1979. R L Wilson

As part of its duty to co-ordinate services in its area the PTE had been in discussion with the North Western Road Car Company, the Stockport-based subsidiary of the state-owned National Bus Company. Some 60 per cent of North Western's local bus mileage was in the Selnec area. It was announced in November that Selnec was negotiating to acquire this part of North Western's operations and the 272 vehicles which operated it at a cost of £1.95 million. The takeover was effective from 1st January 1972.

This was quite a milestone, being the first takeover by a PTE of a substantial part of the operations of an NBC subsidiary. Indeed only one other PTE – West Midlands – managed to conclude a similar deal. The others had to settle for operating agreements.

In advance of the takeover, PTE operations staff were eyeing the North Western lowheight Fleetlines for possible future use as omo buses at Leigh, but nothing came of this.

During 1971 Selnec carried 476 million bus passengers – 50 million fewer than had used the region's buses in 1969 – while 25 million used BR services in the area. The PTE employed 10,054 people at the end of the year including 4,055 drivers and 2,095 conductors. The number of conductors had fallen by almost 20 per cent in 12 months as omo was expanded. The drive to extend omo led to one most unlikely bus being fitted with a farebox: ex-Leigh AEC Renown 6905. Like most other operators who tried running half-cab double-deckers without a conductor, Selnec found the conversion less than ideal and abandoned any thoughts it might have entertained about running half-cab omo buses. Instead it concentrated on fitting omo equipment to existing rear-engined vehicles.

Top **Still in Bury livery in this 1971 view, single-deck Fleetline 6096 was one of six similar East Lancs-bodied buses delivered in 1969. They were 41-seaters and had Gardner 6LXB engines. They were withdrawn in 1981.** R L Wilson

Centre **The only Bristol REs in the original Selnec fleet were six RESL6Gs with 43-seat Northern Counties bodies which came from SHMD and ran for Selnec Southern. They were withdrawn by GMT in 1980-81.** R Marshall

Left **Bolton's seven 30ft-long Daimler CVG6-30s of 1958 were all withdrawn in 1971-72. 6607, smartly repainted in Selnec livery in this 1971 view, was among the last to go. It had a 74-seat East Lancs body, an unusually high capacity for a 1958 bus.** R Marshall

1972: North Western Takeover

The takeover by Selnec of a major part of the North Western Road Car Company was the first significant expansion by acquisition of any PTE. The vehicle for the takeover was a new company, the North Western (Selnec Division) Road Car Company Ltd which initially took responsibility for the transfer of staff, vehicles and premises.

Five of North Western's depots and 272 assorted buses of AEC, Bristol, Daimler, Dennis and Leyland manufacture passed to Selnec. The depots and their allocations were:

Stockport (Charles Street)	132
Urmston	46
Altrincham	42
Oldham (Clegg Street)	30
Glossop	22

Charles Street was the company's headquarters and workshops. The North Western fleet was a modern one and most of the vehicles acquired by Selnec were less than 11 years old. On 4th March the North Western (Selnec Division) Road Car Company Ltd was reformed as the Selnec Cheshire Bus Company. Selnec Cheshire used the standard lazy-S logo in brown, with the fleetname Cheshire, and this was applied to acquired vehicles before they were repainted – not showing up very clearly on the background of North Western's red livery. The Selnec Cheshire fleet retained its North Western fleet numbers. These were broadly in the range 1 to 344 and 721 to 971. None clashed with existing buses in other Selnec companies, although some Fleetlines in the 201-10 range did occupy the same numbers as the surviving Bedford VALs in the Selnec coach fleet.

Early repaints in orange and white included 1961 Dennis Loline 893, 1963 Leyland Leopards 958/60, 1965 Daimler Fleetlines 182/8 and 1970 Bristol RE 330. The oldest bus acquired from North Western, 1957 AEC Reliance 721, was repainted too. The 1963 Leopards, of which there were five with Alexander Y-type bodies, were known as Travelmasters; the previous year's batch of Leopards, with a BET-inspired style of Alexander body, were called Highlanders.

Selnec Cheshire relinquished Knutsford local service 96 in March. It was taken over by Crosville. No vehicles were involved.

The integration of the North Western business into Selnec was the responsibility of Jack Thompson, controller of integrated operations, who had been general manager of Selnec Central. Norman Kay, the former Bury general manager and head of Selnec Southern, succeeded Thompson as Selnec Central's general manager. Richard Cochrane took temporary control of Selnec Southern. Tom Dunstan, North Western's traffic manager, now headed Selnec Cheshire.

Integration was one of the aims of the PTE and one example of this appeared in March when former Salford services 64 and 66, from Piccadilly to Peel Green, were linked with the 219 from Piccadilly to Guide Bridge and Ashton-under-Lyne, which had been a joint Ashton/Manchester service. The new through route was numbered 64/66 and was run jointly by Central and Southern.

A new subsidiary, Selnec Transport Services, was incorporated in April. It was primarily concerned with vehicle leasing – Selnec was an early user of leasing as a method of acquiring new buses because of the tax advantages it could bring.

Selnec's fourth fares revision came into effect in May. It was centred on fares reductions, a rare event in the inflationary 1970s, to eliminate anomalies in ex-North Western fares which were on a higher scale than those of Selnec.

The last outstanding municipal bus order was delivered in 1972 to Selnec Central. This was a batch of 34 Daimler Fleetlines (2271-2304) which had been ordered by Manchester. The body order was originally placed with East Lancs but was switched first to Park Royal and then to Roe because of the disruption caused to East Lancs production by the fire at their factory in 1970. These were the only Mancunian-type bodies built by Roe. They were also the last Roe bodies to be supplied to the PTE. Roe was, of course, an associate of Park Royal within the British Leyland group.

The balance of a Rochdale order for Swifts and a Bolton order for Atlanteans, delivery of which started in 1971, was completed early in 1972. The Atlanteans had the last East Lancs bodies to be built for the PTE.

The EX experimental series, started with the six prototype standard Atlanteans in 1970-71, was expanded in 1972. Further prototype standards, EX7-21, were delivered. These again had Northern Counties bodies but this time on Daimler Fleetline CRG6LXB chassis. The last five were of dual-door layout and seated 72, compared with 75 on the single-door buses. As with EX1-6, all operated with Selnec rather than divisional fleetnames. Unlike EX1-6 they had their front service numbers displayed in the near-side aperture, moving away from established Manchester practice and setting a new standard for the PTE.

Opposite A minority type boosted by the North Western takeover was the Leyland Tiger Cub. Selnec had started off with 27. This had fallen to 16 running with the Northern and Southern companies, when 14 1960 specimens were acquired from North Western. These all had 43-seat Willowbrook bodies and they were withdrawn in 1972-73. Of these, 795 heads into Stockport in July 1972 with the Cheshire flash above the front wheelarch. R Marshall

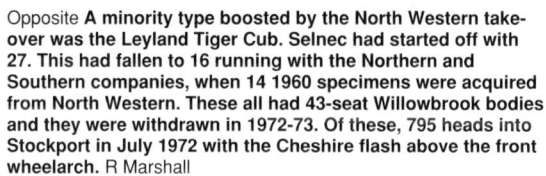

Above left The oldest double-deckers taken over from North Western were 15 Dennis Loline Is which had East Lancs bodies with forward entrances and had been new in 1960. These were withdrawn in 1972-73 without receiving Selnec livery. On 812 the Cheshire flash has been applied over the roughly obliterated North Western fleetname. R Marshall

Above right The most modern North Western double-deckers were Daimler Fleetlines with lowheight Alexander bodies. In 1972 Selnec Cheshire 13 received a white and blue overall advertising livery for greyhound racing at Belle Vue. It brought to four the number of overall advertising buses in the fleet. G R Mills

Centre From 1961 most of North Western's new buses had had Alexander bodywork. The Y-type body on this 1963 Leyland Leopard wears its age well as it takes a turn on the Trans-Lancs Express from Stockport in the autumn of 1972. Five buses of this type were taken over by Selnec Cheshire. They were withdrawn in 1975-76. R Marshall

Left The last Rochdale order to be delivered to Selnec Northern comprised ten AEC Swifts with Pennine bodies. These entered service in 1971-72 and were Selnec's only new AEC buses – although a few Reliance coaches were delivered later in the decade. They brought to 14 the number of Swifts in the fleet. R L Wilson

35

Delivery of the 21 EX-class prototype standard double-deckers was completed in 1972. EX16 was based on a Daimler Fleetline chassis which had been ordered by Bury. The 75-seat Northern Counties body has the route number display in the nearside rather than the offside box; this was to be the future standard. The two-step entrance is clearly visible, as are the two translucent panels in the upper deck ceiling. The front one was positioned to provide maximum illumination for the stairwell. Legal ownership of EX1-21 was attributed to Selnec PTE rather than to one of the companies and they all carried Selnec rather than company fleetnames. *R Marshall*

The last five of the prototype standards had dual-door bodywork, a layout still in vogue in 1972. The staircase was moved rearwards (as was the leading translucent ceiling panel) but otherwise the Northern Counties body differed little from the single-door version. The Fleetline chassis for these buses came from an outstanding Rochdale order. In developing the standard body Selnec's engineers had included a very shallow skirt panel designed to cut accident repair costs. The efficacy of this arrangement appears to have been borne out by EX20 on which the length of skirt panel between the centre exit and the rear wheelarch is missing, presumably after an encounter with a high kerb. EX20 entered service in Barclaycard advertising livery and was the first overall advert in the fleet. *G R Mills*

The 21 EX-class double-deckers were divided between the three divisions. Central had Atlanteans EX1 and EX2, allocated to Queens Road, and the five dual-door Fleetlines, EX17-21 which were sent to Hyde Road. EX17-21 were part of a Rochdale order and in May Mancunians 2266-70 were moved from Hyde Road to Rochdale, where they ran for the rest of the year as temporary replacements. Southern had Atlanteans EX3 and EX4 and Fleetlines EX12-16. The Fleetlines were initially allocated to Ashton (EX12/3), Oldham (EX14) and Stockport (EX15/6). Northern received Atlanteans EX5 and EX6 and Fleetlines EX7-11. Here the Fleetlines went to Rochdale (EX7/8), Bolton (EX9,10) and Bury (EX11). Central's EX20 entered service in Barclaycard livery and was the first overall-advert bus in the fleet.

At the end of the year the 21 experimental buses were renumbered into existing ex-municipal series. EX1-6 went to Ashton as 5466-71. EX7-11 and 17-21 became 6245-54 in the Rochdale series while EX12-16 became 6395-9 in the Bury series, leaving 6394 blank for no apparent reason. These were the fleets which had originally ordered the chassis and with the exception of Barclaycard advert bus EX20, which remained for a period with Selnec Central, the EX-class double-deckers were thus re-allocated to their intended fleets. On renumbering their Selnec flashes were replaced by the appropriate divisional flash. Some ran briefly with Selnec names and divisional flash colours. On internal computer records EX1-21 were listed as 9001-21 because the computer was programmed to accept only figures, not letters, as fleet numbers.

The eight 11m-long Metro-Scania BR110MH integral single-deckers ordered in 1971 were delivered in the summer of 1972, along with eight 11.3m-long Leyland Nationals. The Nationals were EX30-37 and the Metro-Scanias were EX42-49 with the two batches being registered consecutively TXJ507-22K. All were of two-door layout with the Leyland Nationals seating 46 and the Metro-Scanias 44.

The Metro-Scania was built in Birmingham by MCW using Scania running units. The integral body was based on Scania's CR110 design. The engine was an 11-litre naturally-aspirated Scania CO1 rated at 181bhp and it was linked to a Scania automatic gearbox. The Metro-Scania had air suspension.

The integral Nationals were built at Leyland's new Workington factory, owned jointly by Leyland and NBC, and had the standard Leyland 510 8.2-litre 180bhp turbocharged engine driving through Leyland's Pneumocyclic gearbox – although EX36/7 were unusual in having ZF Hydromedia 2-HP-45 two-speed torque converter automatic gearboxes. This was controlled by a three-position rotary selector switch which limited the driver's choice to forward, neutral or reverse. Like the Metro-Scania, the Leyland National used air suspension. EX30 was the first production National.

Both types were put into service alongside each other to enable Selnec to gather comparative operating experience. EX33-35 and EX42-45 initially entered service with

Selnec Northern in Bolton, where they were operated with Atlanteans on the 38-40 Deane Road group of routes and the 66/67 Great Lever circular. Selnec Central got Nationals EX30-32/6 and Metro-Scanias EX46-49 which entered service from Birchfields Road on route 94 (Southern Cemetery to Levenshulme via Piccadilly) in October – but in November they were off the road for a short time because of an industrial dispute by maintenance workers over the EX class vehicles.

1972 saw delivery of the first EX-class single-deckers. EX30-37 were 11.3m-long dual-door Leyland Nationals. EX33 is seen running for Selnec Northern in Bolton in October. Like the EX-class double-deckers it carries Selnec legal lettering and fleetnames. G R Mills

With the Leyland Nationals came eight 11m Metro-Scanias. These, too, were dual-door buses and seated 44; the Nationals were 46-seaters. EX49 is running for Selnec Central in this October 1972 view. The polished mouldings on the Metro-Scania body forced the first major adjustment in the proportions of Selnec's livery with a much greater area of orange than on any previous buses. Manchester City Transport lives on in the bus stop flag. G R Mills

There was a major upgrading of the coach fleet in 1972. A total of 21 new vehicles (235-255) was purchased: eight Bedford YRQ/Duple, five Bedford VAS5/Duple (marking a return to the bigger engine option), two Seddon Pennine T6/Plaxton, two Leyland Leopard PSU5/Plaxton and four Leopard PSU3/Duple. The Leopards – the first in the coach fleet – all had Pneumocyclic gearboxes, providing commonality with Atlanteans. The use of Duple bodywork was new; the coaches delivered in 1970 and 1971 had all been Plaxton-bodied. The two PSU5 Leopards were the first 12m-long coaches in the fleet. At the opposite end of the scale a petrol-engined Ford Transit with 12-seat Deansgate coach conversion was also purchased and numbered 201; it was soon renumbered 215. It was Selnec's only petrol-powered psv.

The Bedford YRQ had a mid-mounted vertical engine, the 150bhp 466 cu in unit already in use in the fleet's newest VALs. The YRQs were Selnec's last Bedford coaches. The Seddons were also the last coaches to be supplied by that company, although Seddon was briefly to play a significant role in PTE bus developments. Unlike previous Seddon coach deliveries the 1972 vehicles had Perkins in-line six-cylinder turbocharged engines rated at 155bhp. The earlier coaches had been powered by 170bhp Perkins V8 engines.

This influx of new vehicles replaced the ex-Manchester Bedford VALs, although before withdrawal three spent a few months in the Cheshire fleet and were given A suffixes to their fleet numbers – 207A, 211A and 212A – to avoid confusion with Daimler Fleetlines numbered in the low 200s.

Transfers of note included ex-Bury single-deck Fleetline 6091 which went to Ramsbottom in January. In early summer similar buses 6092-97 were transferred from Northern (Bury) to Central (Salford) to replace the troublesome ex-Manchester Leyland Panthers. In exchange Northern received ex-Manchester 1965 Atlanteans 3721-27, which operated in Bury, Bolton and Rochdale before being returned to Central at the start of 1973.

To extend one-man-operation in Salford, principally on routes 51 (Victoria to Unsworth) and 73 (Victoria to Whitefield), 11 former Manchester 38xx Atlanteans and four 47xx Fleetlines were moved to Frederick Road. In exchange ex-Salford 1965 Atlanteans 3066-74, which were not equipped for omo, were moved to Hyde Road and Birchfields Road. Similar buses 3055-65 were already running from Manchester's Queens Road garage.

In the summer two Cheshire AEC Reliances, 721/9, were transferred to Selnec Southern at Stalybridge – a surprising move since Stalybridge had no AECs. A Rochdale AEC Swift, 6036, moved to Leigh in July to replace an ex-Leigh Tiger Cub. More soon followed. Leigh, with its 14 Renowns – and a few ageing Regent IIIs – was one of the few Selnec garages running AECs. More surprising transfers from Rochdale were 1959 Regent Vs 6221/2 which reintroduced AEC double-deck operation to Bury – but only for a short time. Ex-Stockport Titans were on the move again in 1972 with three of the unusual Longwell Green-bodied PD2s being

New deliveries in 1972 included the first Leyland Leopards for the coach fleet, four of which were 11m-long PSU3 models with 49-seat Duple Viceroy bodies. This was the last year of Viceroy production and these were the last new coaches with opening windows. Legal ownership of these vehicles was vested in the PTE rather than in one of the companies. R Marshall

Former Ramsbottom PD2A 6402 had been repainted shortly before the formation of Selnec and when withdrawn in 1973 was still in Ramsbottom livery. New in 1962 it had a 63-seat East Lancs body. R L Wilson

All of the later Ramsbottom Titans received Selnec livery. 6403 was the undertaking's first forward-entrance example and again had East Lancs bodywork with 63 seats. It was withdrawn in 1976. Ramsbottom's buses were the only double-deckers in the Selnec fleet not to have route number displays, although these were later added to some. This PD2A is running on an ex-Bury service. R L Wilson

Alexander bodywork was a comparatively rare sight in England when Bury went to the Scottish coachbuilder for bodies on a pair of AEC Reliances in 1964. They had Alexander's stylish Y-type body with 43 seats and were the only Alexander single-deck bodies in the Selnec fleet until the takeover of North Western. They lasted until 1977. R Marshall

Upper centre Selnec's smallest bus was this 21-seat Bedford J2 with Duple Midland body. It came with the Bury fleet and was retained until 1976. In this 1974 view at Ramsbottom it carries Selnec fleetnames after the decision had been made to drop the company names. The legal owner is described as Selnec PTE, but still with the Northern company address. R Marshall

Above Selnec's two Leyland Royal Tigers, both ex-Bolton buses, were withdrawn in 1972. East Lancs bodywork was fitted to 6009, new in 1955. It was the oldest single-deck bus operated by Selnec. R Marshall

sent to Stalybridge where they were later joined by six East Lancs-bodied examples. Six of Oldham's newest Atlanteans, ABU-J-registered 5194-99, were moved to Bury (five) and Stockport (one).

Ramsbottom was converted to total one-man operation in September and its front-engined Leyland Titans, 6401-11, were redeployed in Bury. Its omo fleet comprised seven REN-registered ex-Bury Atlanteans of 1963 which were older than the Titans they were replacing; three 1971 TWH-K-registered Atlanteans ordered by Bolton; plus three single-deckers – ex-Bolton Royal Tiger 6009, ex-Bury Bedford J2 6081, and the Albion Nimbus, 6082. In October all of Bury's omo services were converted to Bell Punch Solomatic operation.

The Royal Tiger's stay at Ramsbottom was short-lived. Selnec only operated two Royal Tigers, former Bolton 6009/10, and both were withdrawn by the end of the year. Two other ex-Bolton single-deckers, 1964 Leopards 6054/5, were also moved to Bury and stayed there until 1975.

Fleet numbers and company allocations for the first of the new standard double-deckers developed from the EX-series were announced in the summer. They were to start a new block of numbers from 7001 with, surprisingly, both Leyland Atlanteans and Daimler Fleetlines in the same number sequence. The Manchester practice of using separate number series for different makes of chassis had had much to commend it; Selnec's move away from this practice was yet further proof – if any were needed – that the PTE was not simply following the norms set by its biggest constituent. The Fleetlines were CRG6LXB models with Gardner engines; the Atlanteans were of the improved AN68 model – easily recognised by its revised three-piece engine cover.

Numerically, the first 150 vehicles were Atlanteans, to be divided between all three companies. Park Royal was to body 145; Northern Counties the remaining five. Selnec Central was allotted Atlanteans 7001-71 and the five Northern Counties-bodied buses, 7146-50, and these were registered VNB101-76L at the Manchester licensing office. Selnec Northern got 7072-7109, registered WBN950-87L in Bolton. Selnec Southern was allocated 7110-45, registered XJA501-36L in Stockport.

As can be seen, each company was responsible for the licensing of its own buses, although all were allocated fleet numbers in a common Selnec series. No attempt was made to match fleet and registration numbers, a practice which had been followed by most of Selnec's constituents and had initially been respected by Selnec in allocating sequential registration numbers to batches of municipally-ordered buses delivered after the PTE's formation.

Selnec Northern 6316 was transferred from Bury to Ramsbottom for the depot's omo conversion in 1972. It is seen in Rawtenstall after the 1974 route renumbering, running as a one-man bus as shown by the fare box logos alongside the destination display and underlined by a tiny 'Pay as you enter' sign in the nearside windscreen. This was one of 15 Atlanteans delivered to Bury in 1963 – the undertaking's first rear-engined buses – with Liverpool-style Metro-Cammell bodies. R Marshall

39

Delivery of the first 150 Atlanteans got under way in August, and with some rapidity. By the end of the year 118 had been delivered, including the five from Northern Counties. The balance followed in 1973.

The order placed for 350 Daimler Fleetlines in 1971 was allocated fleet numbers 7151-7500 and the allocation of buses up to 7328, the expected 1972-73 delivery, was also announced in the summer. Central got 7206-79 which were allocated registrations VNB177-250L. Delivery carried on beyond August 1973 and the VNB-L series finished at 7270 (VNB241L); the remaining nine vehicles had M-suffix registrations. 7206 was exhibited at the 1972 Commercial Motor Show where it carried Selnec, rather than Central, fleetnames.

Northern was allocated two batches of Fleetlines, 7151-86 and 7280-99. These were registered WBN988-999L and WWH21-64L. Southern got 7187-7205 and 7300-28. These were to have been registered XJA537-84L but the series finished at XJA580L and the last four vehicles had M-suffix registrations. For all three companies the Fleetline registrations continued the series started by the Atlanteans.

Park Royal bodies were fitted to 7151-7205 (divided between Northern and Southern), while all of the remainder had Northern Counties bodies. Central's 7206-51 and Northern's 7280/1 were of 72-seat dual-door layout – these 48 were the only dual-door standards (apart from five of the prototypes), although the original plan had envisaged that 229 buses (7206-7434) would have two-doors. All subsequent Selnec (and Greater Manchester) double-deckers were of single-door layout. Fleetline deliveries were slower than those of Atlanteans and only 20, all dual-door Northern Counties buses, were in operation by the end of the year.

There were many detail differences between the Park Royal and Northern Counties standards. The Park Royal body had a more rounded front dome with two ventilation louvres and slightly recessed upper deck windscreens. The Northern Counties dome featured a centrally located grille above flush-fitting windows. The orange band above the lower deck windscreens was deeper on the Park Royal body and the corner pillars were angled back slightly. The panel above the entrance door on the Park Royal body had a slight step up; on Northern Counties bodies it was carried forward at the same depth as the orange band above the lower deck side windows and concealed the top edges of the door leaves.

The first standard Atlantean, Central's 7001, was – not surprisingly – allocated to Hyde Road garage which was adjacent to the Central Works and the Selnec head office in Devonshire Street North. It arrived in August. But Central's 7002-16/8-20 and 7146-50 were loaned to Southern at Oldham, Stalybridge and Stockport when new to cover late deliveries of Fleetlines and allow omo conversions to go ahead. Thus Central garages had to wait until October to see any more new standards, with the delivery of 7021-33 to the ex-Salford garages at Weaste and Frederick Road. Missing bus 7017 entered service at Hyde Road in October as an all-over advert for Berni Inns.

The new standard double-deck series commenced at 7001 and numerically first were Leyland Atlanteans with Park Royal bodies of which 113 were delivered in 1972 and divided between each of the three companies. 7012 was a Selnec Central bus but is seen in Stockport in October 1972 soon after entering service on loan to Selnec Southern. *R L Wilson*

Northern Counties delivered only 25 standards in 1972 and all were of two-door layout. Most went to Selnec Central but two, 7280/81, were allocated to Selnec Northern at Rochdale. On dual-door standards the staircase was moved half a bay to the rear and the seating capacity reduced from 75 (43 up, 32 down) to 72 (45 up, 27 down). The intermediate destination display has been partly blanked off to accept standard Rochdale Corporation blinds. *R Marshall*

Allocation of the first 328 standards

	Central	Northern	Southern	Total
Atlantean/Park Royal one-door	71	38	36	145
Atlantean/N Counties one-door	5	–	–	5
Fleetline/Park Royal one-door	–	36	19	55
Fleetline/N Counties two-door	46	2	–	48
Fleetline/N Counties one-door	28	18	29	75
Totals	150	94	84	328

Most of the buses loaned to Southern had found their way back to Central by the end of the year, although a few remained at Stalybridge until March 1973. The redistribution of the 23 Central buses returned from Southern meant that there were standard Atlanteans at the former Manchester garages at Hyde Road, Northenden, Queens Road and Princess Road by the end of the year. Birchfields Road had to wait until January 1973 for its first standards.

Central's first standard Fleetlines arrived in October and were allocated initially to Frederick Road (five) and Hyde Road (12), with one going to Northenden in December, by which time eighteen, 7206-23, had been delivered. No great effort was made by Central to keep standard Fleetlines and Atlanteans at different garages.

All of Northern's first standard Atlanteans, 7072-7109, went to Bolton with delivery extending from August through to January 1973. The only standard Fleetlines delivered to Northern in 1972 were dual-door 7280/1 which were allocated to Rochdale.

Delivery of Southern Atlanteans started in August with the majority going to Stockport although small numbers were also allocated to Oldham, Stalybridge and Ashton.

As new standards arrived, so elderly acquired vehicles departed. Withdrawals included large numbers of ex-Manchester Daimler CVG6s in the 44xx series, new in the mid-1950s, although the last of these actually survived until 1974. A fair number of PD2s from the same era, in the 34xx series, also left the fleet. Rather more surprisingly, a start was made in November on withdrawing Manchester's first Atlanteans, 3621-30 of 1960, and all ten had gone by 1973 to be outlasted by older conventional front-engined buses. The high cost of re-certification after 12 years' operation was a major factor in the decision to dispose of the Atlanteans.

New buses were needed to replace some of the PTE's ageing inheritance, such as 19-year-old Daimler CVG6 4449, still running in Manchester red in October 1972 – albeit with Selnec-style Helvetica fleet number transfers on the centre of the bonnet. This Metro-Cammell-bodied bus was withdrawn in 1973 but the last of the batch ran until 1974. *G R Mills*

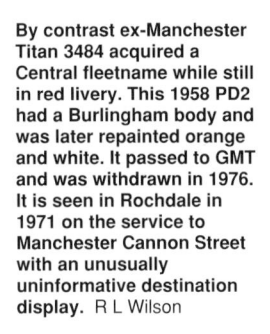

By contrast ex-Manchester Titan 3484 acquired a Central fleetname while still in red livery. This 1958 PD2 had a Burlingham body and was later repainted orange and white. It passed to GMT and was withdrawn in 1976. It is seen in Rochdale in 1971 on the service to Manchester Cannon Street with an unusually uninformative destination display. *R L Wilson*

41

A number of non-standard types vanished in 1972. In Selnec Southern the last of the five Atkinsons acquired from SHMD, 5068, left the fleet; it had been Selnec's only centre-entrance single-decker. No Atkinsons received Selnec livery. Other SHMD departures were the two manual-gearbox Daimler CSG6s, 5696/7, and the last of the batch of six centre-entrance CVG6s, 5671-76. Centre-entrance buses had, understandably, no part to play in a forward-looking transport business and none were repainted in Selnec colours. They were, of course, old too.

A start was made on disposing of the large fleet of Selnec Northern AEC Regent Vs acquired from Rochdale – but the last of these was not to vanish until 1977. The three surviving ex-Bury BEN-registered PD2 Titans – new in 1953 – were all withdrawn by the end of the year. Two premature withdrawals – because of fire damage – were ex-Manchester Fleetlines 2073 (a Mancunian) and 4730.

Four Seddon buses of a new design were ordered in 1972 and one, numbered EX58, was completed in time for the Commercial Motor Show at Earls Court in September. The Seddons were an attempt to explore the opportunity to use small buses and were 25 seaters with room for up to 10 standees. The chassis, the Pennine IV:236, had a 9ft 6in wheelbase and the Seddon-built body was only 7ft 6in wide. The buses were just under 21ft long, weighed 4 tons 5 cwt and had single rear wheels, which led to unusual handling characteristics, especially on wet surfaces.

The small Pennine was derived from Seddon's full-size chassis but used a Perkins 4.236 3.86-litre four-cylinder engine instead of the six-cylinder unit used in larger Pennine models. This was rated at 80bhp at 2,800rpm. A Turner five-speed synchromesh gearbox was fitted and the chassis was air-braked. The little Pennines had smaller (8.25 x 20) wheels and tyres and 12 volt electrics. All four were delivered in 1972 with three going to Southern at Stockport (EX56-58) and one going to Northern at Ramsbottom (EX59) where it replaced Bedford J2 6081.

An elderly addition to the operational fleet in 1972 was ex-Stockport 295, a 1951 Leyland Titan PD2 which had had its roof removed while still in Stockport ownership and was used for tree lopping. This was brought back up to psv standard and made available for private hires and promotional work. It was repainted into Selnec livery early in 1973, becoming the oldest bus to carry fleet colours.

Repainting of acquired vehicles continued, but many buses were still in their original liveries. From the end of 1972 those Selnec Southern buses still in municipal colours had the Selnec logo and company fleetname applied on the side, generally on a white background. By contrast Selnec Northern did not apply its logo to any buses in municipal liveries.

Below left The arrival of new buses enabled the oldest surviving ex-Bolton Titans to be withdrawn in 1972. These were 1956 PD2s with Metro-Cammell Orion bodies. 6577 managed to retain its Bolton Transport fleetname and municipal crest until the end. R L Wilson

Below right Another group of Bolton buses to vanish in 1972 were the 1957 Daimler CVG6s with East Lancs bodies. 6591 was an early Selnec repaint. It passes a new Skoda Octavia estate car. R L Wilson

Below left Withdrawal of ex-Rochdale AEC Regent Vs started in 1972. 6220 with 61-seat Weymann body was one of four delivered in 1959 which were the only ones to be fitted with platform doors. This bus had an AEC engine; most of the ex-Rochdale Regents were Gardner-powered. R L Wilson

Below right Eight 1953 Weymann-bodied PD2s were acquired from Bury and operated until 1971-72 before being withdrawn, still in Bury livery. G R Mills

New at the 1972 Commercial Motor Show was the Seddon Pennine IV:236, a midibus developed for Selnec to test the viability of small buses. It was exhibited on Seddon's stand with Selnec fleetnames and coach-style wheeltrims on the front wheels – which were removed before the bus entered service. It was one of four similar buses with 25-seat Seddon bodies with shallow windows. *Stewart J Brown*

Stockport 295 was a 1951 all-Leyland PD2 which had been used as a tree-lopping vehicle by both Stockport and Selnec. In 1972 it was re-certified to psv standards and added to the Selnec Southern operational fleet as an open-top bus for promotional use. It is seen in Stockport depot after being repainted orange and white. *R Marshall*

Interest in the use of two-way radios to improve service quality was growing in the early 1970s and Selnec Northern ordered 200 bus radio sets in 1972, half of which were to be used to equip the Bury fleet.

Rail developments in 1972 included good and bad news. Among the good news was the granting in August of Royal Assent to the Bill allowing construction of the central Manchester underground rail link, the Picc-Vic line. Manchester was at last going to get an underground railway! Preliminary survey work had in fact started in April. Linked with this was approval for a new bus-rail interchange at Bolton. Interchanges were also under consideration for Bury and Altrincham. A study of rail services to the latter envisaged the introduction of a peak-hour service with a five minute frequency and bus-rail interchanges at major stations on the line.

The bad news was the closure of the Bury –Ramsbottom–Rawtenstall line. Passenger services over the 8.25 mile route, operated by diesel multiple units, ceased on 4th June. A replacement bus service, normally run by Albion Nimbus 6082 – which might say something about the level of patronage for the rail service – was introduced from Bury to Rawtenstall via Summerseat and numbered 8. Every cloud is said to have a silver lining and the Bury line was later to be re-opened as a successful tourist attraction by the East Lancashire Railway, running preserved steam locomotives.

There was news in the meantime of a reprieve for the line from Oldham to Rochdale and refusal of permission for the closure of the Stockport to Stalybridge line, which had earlier escaped the Beeching axe in the mid-1960s. This became the first service in the Manchester conurbation to be operated by pay-trains, on which the guard performed the role of a bus conductor, collecting fares and issuing tickets. Electrification of the Hazel Grove line was recommended in a review of the rail network in the Selnec area.

In November Selnec became the first PTE to conclude an operating agreement with British Rail, under which the PTE became responsible for 25 local services on 19 rail lines, the largest local rail system in Britain

44

Opposite **Freshly repainted in 1972, ex-Salford Atlantean 3073 heads along Cannon Street, Manchester. It has a 77-seat Metro-Cammell body. After buying this batch of Atlanteans in 1965, Salford reverted to Titans for its 1966-67 deliveries. In pre-PTE days the 96 service had been a joint operation between the Salford and Manchester undertakings.** R Marshall

Opposite lower **Withdrawal of the unreliable ex-Manchester 1967 Leyland Panthers, of which there were 29, started in 1971. However a few received Selnec Central livery. No.54, passing through Weaste on an ex-Salford service, was one and was also among the last of the ex-Manchester Panthers to be withdrawn in 1974. Metro-Cammell built the 44-seat body.** R L Wilson

Left **Among the more unusual vehicles acquired from Oldham were two Leyland Tigers Cubs with dual-door Pennine bodies. New in 1965, they were long-lived for lightweight vehicles and survived until 1978.** R Marshall

Below **Other Tiger Cubs in the Selnec Southern fleet which also enjoyed long lives were four ex-Stockport buses with 44-seat Crossley bodies. New in 1958, they lasted an impressive 17 years, being taken out of service in 1975.** R L Wilson

outside London according to the PTE, and one that was losing around £5 million a year. The agreement was backdated to January.

This was all in accordance with the requirements of section 20 of the 1968 Transport Act, under which the PTE had to accept increasing financial responsibility for local rail services in the Selnec area. In 1972 the rail network cost £6.9 million to run – but earned fares revenue of only £2.7 million. In 1972 the PTE became responsible for funding 10 per cent of the subsidy to the 21 grant-aided services in its area, a figure which was increased to 20 per cent in 1973 and 30 per cent in 1974 when the situation was to be reviewed in the light of impending local government reorganisation.

There were 113 stations in the Selnec area, three of which had catered for fewer than 100 passengers a day in 1971: Westhoughton (36 passengers), Royton Junction (92) and Reddish South (87). The busiest station outside central Manchester was Altrincham with 8,837 passengers daily. This was followed by Stockport (7,211) and Bolton (4,939).

The Selnec Parcels operation received a boost with the opening of a new depot in the former Parrs Wood bus garage. This handled 8,000 parcels a day, with a fleet of 40 vans.

The year had opened with the absorption of much of the North Western company's operation. It closed with the first hint of another significant expansion. In December an agreement was signed with Lancashire United Transport which gave the PTE still greater control over LUT's services – but more importantly also gave it the option to acquire the company on or after 1st January 1976. There had in fact already been an important change in the constitution of LUT during the year with the formation of a new holding company, Lanaten, which owned LUT and in which the PTE had invested £2.6 million – the sum it would pay if it chose to acquire LUT in 1976.

During 1972 the enlarged Selnec operation employed 10,733 people, operated 2,610 vehicles, carried 523 million bus passengers and made a profit of £1.4 million. Local rail services carried 22 million people.

45

1973: Reorganised Again

The year started with the publication of Selnec's public transport plan for the future, Lifeline 2000. It envisaged completion of the 2.18-mile long Picc-Vic tunnel by 1978, creating a 50-mile rail system with 35 stations which would link Bolton and Bury on the north, via the Bury line to Alderley Edge on the south (via Styal and Stockport) with links to Macclesfield and Hazel Grove. The report called for bus priorities on 15 radial routes to central Manchester and highlighted the need for improved fare collection systems and the extension of radio control. It pointed to the need to renumber routes to avoid duplication and said that the use of route suffix letters would be greatly reduced.

On the vehicle front it confirmed that the fleet would standardise on 9.5m double-deck buses, 10m single-deck buses and 11m coaches with a 12-13 year life. And, in a move which was later to seem to have been ahead of its time, noted that small buses of 20-25 seat capacity offered potential for developing routes in town centres, residential estates and rural areas.

Rapid transit, the great white hope of 1960s planners, also had a part to play with the revival of a plan for a line serving the University, Withington and Northenden which could be linked through the city centre to run on to Salford and Eccles. Also proposed was a connection from Cheetham Hill/Broughton to Market Street. For Wythenshawe the report envisaged reserved bus lanes with stations at half-mile intervals. Sadly, none of this was to be.

What was to be, was the first bus lane on a trunk route in Manchester, a 0.75-mile stretch in Oxford Road, which came into operation on 26th February. The results it achieved were inconclusive – but encouraging enough for the PTE to examine the wider use of bus lanes.

Above **The last two-door standards, on Atlantean chassis, were delivered from Northern Counties in 1973. This view, in Salford's Victoria bus station, allows comparison between the Northern Counties and Park Royal standard bodies; the bus on the left, 7021, is a Park Royal-bodied Fleetline. The company fleetnames were abandoned later in 1973 and 7021 carries Selnec on the side. However this is in fact a summer 1975 view – both buses carry GMT depot allocations (Weaste) alongside their fleet numbers, and have GMPTE legal lettering.** R Marshall

Upper left **On unrepainted buses at Oldham the Selnec fleetname was applied on a white background which was untidy but at least visible. 1957 Roe-bodied Titan 5294 is seen in Oldham town centre in 1973: it was withdrawn in 1974.** R Marshall

Left **Northern Counties delivered 120 standard bodies on Daimler Fleetline chassis in 1973. Among those allocated to Rochdale was 7285.** R L Wilson

Above **The new standards spread quickly. In July 1973 two Selnec Northern Park Royal-bodied Fleetlines in Bury wait behind an East Lancs-bodied Fleetline which had been new to the town's municipal fleet in 1968.** R Marshall

Below **The arrival of new buses in Bury allowed the withdrawal of the ex-municipal Leyland Titan PD3s, 25 of which had been delivered in 1958-59. They had 73-seat Weymann bodies with the unusual refinement, for a rear-entrance municipal bus, of platform doors. Withdrawal was spread over the period 1971-75.** R Marshall

Bottom **An order for 25 Bristol VRTs with Eastern Coach Works bodies had been placed by North Western before the Selnec takeover. These were delivered in 1973 and had Selnec-style destination displays and interior trim in an otherwise standard National Bus Company specification lowheight ECW body. Early deliveries carried Cheshire fleetnames and were the only new buses to do so.** R Marshall

Deliveries of standard double-deckers continued apace in 1973. The 32 outstanding vehicles from the first 145 Park Royal-bodied Atlanteans were accompanied by the 55 Park Royal-bodied Fleetlines (7151-7205), bringing to an end Park Royal's contract to supply 200 bodies. More Northern Counties-bodied Fleetlines were delivered – 132 during 1973 – taking the fleet numbers up to 7358. Fleetline 7292 had an experimental gearbox control manufactured by Sevcon of Gateshead and did not enter service with Selnec until 1974. An order was placed with Leyland for a further 179 Atlanteans and 100 Fleetlines, and the allocation of the balance of the original order for 500 buses was made:

7329-7434	Central
7435-7466	Northern
7467-7500	Southern

The new Fleetlines delivered to Northern in 1973 went to Rochdale, Bury and Bolton – at the last-named introducing Daimler Fleetlines to a predominantly Atlantean fleet. Those for Central were divided between all of the company's seven depots. Southern's went to Stalybridge, Oldham and Ashton. The last four buses of the original Southern allocation, 7325-28, arrived after the June restructuring which saw the dissolution of the companies and three were allocated to Central division's Princess Road.

Double-deck variety was introduced by a batch of 25 Bristol VRTSL6Gs which had been ordered by North Western before its demise. These had Gardner 6LXB engines and standard NBC-style lowheight Eastern Coach Works bodies but with Selnec interior trim and destination displays. Numbered 400-24 with matching registration numbers AJA400-24L they were allocated to Cheshire, although only 400-15 were delivered before Cheshire was dissolved on 21st May and its operations merged with the Southern and Central divisions. The VRTs were the only new buses delivered with Cheshire fleetnames and the accompanying brown Selnec flash.

47

The dissolution of Selnec Cheshire allowed the closure of the former North Western garages at Urmston, Oldham (Clegg Street) and – briefly – part of the Stockport (Charles Street) garage. There were no redundancies.

Central took over Cheshire's Altrincham depot (with 43 buses) and in addition acquired 42 Cheshire vehicles which were moved from Urmston into Princess Road depot. Before Urmston had closed, ex-Manchester PD2 3473 and CVG6 4554 were transferred in to release unfamiliar ex-North Western types for crew training at Princess Road. The types transferred into the former Manchester depot were 34 lowheight Fleetlines and eight Bristol RELLs, all bodied by Alexander. The remainder of the Cheshire fleet – around 160 buses at Stockport, Oldham and Glossop – went to Southern.

With the disappearance of Selnec Cheshire it was not long before ex-Manchester buses appeared at erstwhile North Western depots. Fourteen Mancunian-style Fleetlines were transferred to Altrincham in the summer along with some workings on routes 263/4 from central Manchester.

Most of the former Selnec Cheshire services were renumbered at the same time and two passed to Crosville: 92 Wilmslow to Knutsford and 218 Altrincham to Tatton Park. Much of their mileage was outside the PTE area and Selnec Cheshire had retained them in 1972 because it had surplus staff and could provide the routes more economically than could Crosville. This arrangement ceased with the dissolution of Selnec Cheshire.

The confusingly-initialled Lower Mosley Street bus station in Manchester (often abbreviated to LMS, initials more usually associated with tracked transport) closed on 13th May. Prior to its closure route 6, to Glossop, had its city terminus moved to Whitworth Street West, outside Deansgate Station while the 32 to Middlewood was transferred to Chorlton Street coach station.

Short Leyland Nationals and Metro-Scanias, four of each, joined the long versions delivered in 1972. These were generally similar to the 1972 buses but had 40 seats. They were again of dual-door layout. They were numbered EX38-41 (Nationals) and EX50-53 (Metro-Scanias) with registrations VVM601-608L and were allocated to Stalybridge, entering service in June.

One further Metro-Scania was purchased. It was another 40-seater (EX60, VVM609L) but it differed from the earlier vehicles in having an encapsulated engine compartment to minimise exterior noise. This it did effectively and was marketed as the Hush Bus. Its drive-by noise level was in the region of 77dB(A) compared with around 88dB(A) for conventional buses in the fleet. It went into service in July on route 203 from Manchester (Victoria Station) to Reddish.

National EX38 had unusual GKN-SRM transmission. EX39, which had a ZF gearbox, was loaned to Sheffield Transport for two weeks in May. It also visited Plymouth.

The deliveries of Nationals, Metro-Scanias and Seddon Pennines in 1972-73 left two blank numbers in the single-deck EX series which had started at EX30. These numbers were filled in May by the delivery of two most unusual vehicles – Northern Counties-bodied Mercedes-Benz O.305s. The O.305 was Mercedes' standard rear-engined city bus and the two for Selnec, EX54/5, were the only ones ever to run in Britain. The Northern Counties dual-door 43-seat body was integrated with the O.305's air-suspended tubular-steel underframe. The O.305 was powered by an 11-litre Mercedes OM407h rated at 210bhp and driving through a Mercedes automatic gearbox. In deference to local political sensitivities neither bus carried a Mercedes badge. They went into service at Atherton, running on hire to LUT on service 83 to Mosley Common, at the end of the summer. The Mercedes did not offer any significant advantages over the Leyland National and no more were contemplated.

Top **Lower Mosley Street bus station in Manchester closed in 1973. An ex-Manchester Burlingham-bodied PD2 waits to leave on the Glossop limited stop service in 1972. Until the North Western takeover this was the only Selnec service to use Lower Mosley Street.** M Fowler

Above **Short Metro-Scania EX60 was delivered in 1973 and had an encapsulated engine compartment to reduce noise levels. It was promoted as the Hush Bus and is seen operating in Bolton in 1974. It brought to 13 the number of Metro-Scanias in the Selnec fleet, all of two-door layout.** R L Wilson

Top The Silent Rider battery bus was built in 1973 at a cost approaching £100,000 which would have bought five diesel-powered double-deckers. It is seen in London, on demonstration to the technical press. Alan Millar collection

Above One of the Department of Trade & Industry's unsuccessful Crompton Electricars battery-electric buses was used on the new Bolton town centre circular service in the spring of 1973. Allocated fleet number EX100 and carrying the Selnec Northern logo, it was a 19-seater with a Willowbrook body on a modified British Leyland 990FG truck chassis. R Marshall

The costs and long-term availability of fossil fuels were subjects of discussion in the early 1970s. Against this background, and concern about atmospheric pollution and high maintenance costs, Selnec pursued another single-deck development which was to prove fruitless: the Seddon-Chloride battery bus. This, like all other battery-powered buses, suffered from the twin problems of range and weight. The 330 volt lead-acid traction battery had a limited range (about 40 miles) – but despite that the bus still weighed over 50 per cent more than its diesel-powered equivalent. At an incredible 12 tons 17 cwt 2 qr it was the heaviest bus ever to run in Manchester.

Novel features included regenerative braking which harnessed the power of the 72kW traction motor when the bus was decelerating, and a charging system which could fully recharge the batteries in 3½ hours. A Webasto oil-fired heater provided saloon heating.

The Seddon-Chloride bus, named Silent Rider, had thyristor control for smooth acceleration and after trials at the Transport & Road Research Laboratory at Crowthorne, Berkshire, was delivered to Oldham – convenient for the Seddon works – in November. It was certified for passenger service in January 1974 and was then demonstrated to the press in London in March with the aim of putting it into operation in April, although in fact a further 12 months were to pass before it actually entered revenue-earning service. The Silent Rider was capable of speeds up to 40mph and was based on a 10m Seddon Pennine RU chassis with a 43-seat two-door Seddon body – the last two-door bus purchased by Selnec. The chassis was strengthened and the suspension uprated – hardly surprising in the light of its weight. It wore a modified livery with orange relief on the roof only. It was numbered EX61.

After a study of the fleet and operations at Hyde Road garage, Chloride concluded that up to 40 per cent of duties could be covered by battery-powered buses even with their acknowledged limited range. The aim was to have the bus available for operation between 0430 and 1000 and again between 1500 and 2030 with the times in between being used to recharge the batteries. But not only was the Silent Rider heavy and of limited range; it was expensive. A fleet of 20 would have cost around £1.5 million, making them £75,000 each. This sum would have bought 90 standard Atlanteans, then costing around £16,500 apiece. It was clear which was better value, although Selnec was in fact talking to Chloride, the Department of the Environment and the Department of Trade and Industry about the possibilities of running 200 battery buses. The Silent Rider project cost just under £100,000.

Earlier in the year, Selnec had had its first taste of battery buses when it borrowed the Department of Trade and Industry's Crompton Electricars demonstrator, CWO600K, which was allocated fleet number EX100. This was based on a modified British Leyland 990FG light truck chassis and had a Willowbrook body. It was small – 22ft long on a 13ft 4in wheelbase chassis – but heavy, with an unladen weight of 7.75 tons and a gross vehicle weight of 9.5 tons. Capacity was 18 seats and room for eight standees, all that could be managed within the 9.5 ton GVW. Performance, for long the weak point of battery-powered buses, remained a weak point: the top speed was 25mph and its range with 13 stops per mile was only 35 miles, which was considerably less than a single shift for any self-respecting diesel bus.

EX100 joined Seddon Pennine EX59 (transferred from Ramsbottom) on a free Bolton town centre service, launched for an experimental period of three months in February, and between them the two buses demonstrated that however attractive it might be to have silent and pollution-free electric traction, the use of battery power was still not on. EX100 was on loan from March until May and re-appeared for a further short spell in December. The free Bolton service, marketed as Centreline, ran every seven minutes from 0930 to 1730.

49

Another unusual bus to appear on the Bolton town centre service was Edinburgh Corporation 109, a Seddon midi built to what was basically the Selnec design and operated on hire pending delivery of Selnec's own new midis. It ran in full Edinburgh livery but with the addition of Northern fleetnames and was delivered direct from the Seddon factory in Oldham. *R L Wilson*

Electrification might have been out, but metrication was in. From the spring new buses were delivered with the unladen weight inscribed on the side in kilograms rather than in avoirdupois tons, hundredweights and quarters. The first was 7252.

More Seddon Pennine midibuses were on order for 1973 but pending their delivery as 1704-1714 Selnec hired six Seddons from Edinburgh Corporation Transport to cover midibus services. Edinburgh's BWS102/3/5L and CFS106/8/9L ran for Selnec, with the last two going into Selnec service before being delivered to Edinburgh. The first arrived in May. They ran in Edinburgh's dark maroon and white livery but with Selnec divisional logos. Two ran from Bolton (108/9) on the town centre service and two from Frederick Road (103/6) on the Swinton midibus routes.

Fleet numbers 1700-3 were used for the original four Seddon midis, delivered as EX56-59. The 1973 buses appeared with a new style of Seddon bodywork with much bigger windows. The specification was revised and incorporated twin rear wheels to improve handling. The services earmarked for midibuses were Central's 36, operated by Frederick Road and providing an off-peak hourly shoppers' service between Swinton and Pendleton shopping centre, the Bolton Centreline, Bury 8 to Rawtenstall and Southern shoppers' services 373-376 in the Hazel Grove area. The Hazel Grove services took the midis into residential roads not previously used by buses and this provoked some protest. The services were short-lived.

Selnec initially persevered with the route numbers of the operators which it had acquired and was thus in the unhappy position of having no fewer than seven routes numbered 1. These were in Salford, Bolton, Bury, Leigh, Rochdale, Ashton and Oldham. Some minor route renumbering took place but it was not until 1973 that a comprehensive system was evolved, with blocks of numbers for different areas:

1-299 – Manchester, Salford, Urmston, Altrincham
300-399 – Stockport, Ashton, Stalybridge, Glossop
400-499 – Oldham, Rochdale, Bury and Ramsbottom
500-599 – Bolton and Leigh (including some LUT and Ribble services)
600-699 – allocated to routes to be taken over from Wigan and LUT
700-999 – works, schools and other special services.

Implementation was slow. The Stockport group (300-399) was largely renumbered in May, as were the Selnec Cheshire routes. Indeed two new routes introduced in Stockport in November 1972 had been numbered 300/1, presaging the new system. Some Salford works services were renumbered in the 700 series in April. But Oldham did not renumber until December and Bolton, Bury and Leigh until March 1974. Rochdale's renumbering also took place in the spring of 1974. Where possible an attempt was made to reflect something of the old number. Thus former Manchester services 51/2 and 63/4 became 151/2 and 263/4. Many Oldham route numbers simply had 400 added to them. Similarly most Bolton route numbers were increased by 500.

In the autumn proposals were drawn up for some partial fleet renumbering. Coaches would fall into the series 0001-0099 – with Selnec standard specification single-deckers being numbered between 0100 and 0999. It was not intended that the surplus zero would be carried on vehicles; this was purely an administrative change. Other new series planned, and in some cases introduced, were:

1300-1399 new types of single-deckers
1400-1599 new types of double-deckers
1600-1699 Selnec Central single-deckers
1700-1799 midibuses
1800-1899 ex-Selnec Cheshire Dennis Lolines
1900-1999 ex-Selnec Cheshire AEC Renowns
3000-3999 Leyland double-deckers, Selnec Central and Wigan
4000-4800 Daimler double-deckers, Selnec Central and ex-Cheshire

Despite the intake of new buses, this was the period of the three-day working week (because of a national coal miners' strike) and deliveries were in fact behind schedule. This, linked to shortages of spares and a staff turnover of 23 per cent (compared with 14 per cent in 1972), led to problems in maintaining service reliability. In December the PTE announced that because of the fuel crisis no serviceable buses would be disposed of and it set about a limited refurbishment of three ageing NNB-registered 44xx-series Daimler CVG6s.

One result of the closer involvement with LUT was the launch in January of Selnec Travel running a programme of coach tours in Britain and Europe which were operated by LUT, which held the necessary tours licences, using Selnec coaches. Selnec Travel also handled bookings for other operators' tours, including Shearings, Happiway-Spencer, Smiths, Ribble and Florence & Grange. The first continental tour departed in April.

The coach fleet was renumbered in 1973 by the subtraction of 200 from existing fleet numbers. It now fell in the range 13 to 55, numbers previously used for ex-Manchester single-deck buses. New coaches delivered in 1973 were all 11m-long Leyland Leopard PSU3 models with Pneumocyclic gearboxes. Bodywork was by Duple (the new Dominant),

Plaxton (Elite Express III) and, surprisingly, Eastern Coach Works. ECW supplied, for the first time on a Leopard, the coach body design being built for NBC subsidiaries on Bristol RE chassis. There were ten ECW 47-seaters (56-65), three Duple 49-seaters (66-68) and seven Plaxton 49-seaters (69-75) to give a total intake of 20 coaches. In July management of the Selnec private hire operation was centred in Stockport, although the coaches remained with the divisions.

A new divisional structure was implemented in June to replace the three companies. Selnec Central was split into three divisions. Central East was based at Birchfields Road under general manager John Marsh, a former Oldham Corporation traffic superintendent, Central North West at Frederick Road (Salford) under Mike Hicks, who had previously worked in Africa with the BET group, and Central South at Princess Road under Tom Dunstan who had been general manager of Selnec Cheshire. Selnec Northern became the Northern division with four districts at Bolton, Bury, Leigh and Rochdale. It remained under the control of Jim Batty. Similarly Selnec Southern became the Southern division with three districts at Ashton/Stalybridge, Oldham and Stockport. It was headed by Ken Holt, who had been with Manchester City Transport. The three company fleetnames – Central, Northern and Southern – and their coloured flashes were discontinued and as buses were repainted they were given Selnec fleetnames with an orange flash. Each of the divisional general managers reported to Bill Broadbent, chief operations executive, whose management team included former Bury general manager Norman Kay, as services and operations manager, and Brian Holcroft as chief operations engineer. Former Oldham general manager Harry Taylor became chief development engineer. Richard Cochrane assumed the new post of corporate services manager.

Top Ten of the 20 new Leyland Leopards delivered in 1973 had ECW coach bodies – an unusual body choice outside the National Bus Company, and a combination which at that time was unique to Selnec. The body was designed for the rear-engined Bristol RE and within a short time required strengthening around the rear boot area. 59 loads in Rochdale for Bolton on the Trans-Lancs Express. R Marshall

Centre Shortly after the dissolution of the Cheshire company the other three companies were re-organised into a new divisional structure and all repaints were given Selnec fleetnames with orange flashes, as demonstrated by ex-North Western Fleetline 103, seen in Altrincham after the 1973 route renumbering. R Marshall

Right Although the lowheight Alexander body became quite a common sight in the Selnec fleet after the North Western takeover, the highbridge version of the same body was restricted to one batch of 15 Fleetlines delivered to Bury in 1964. These 74-seaters spent their entire lives in Bury; most were withdrawn in 1977 although a few were recertified and lingered on until 1980. To minimise windscreen replacement costs Bury had specified split flat glass windscreens instead of the more attractive single-piece double curvature screen which had been designed for the body. R Marshall

There were six Leyland Panthers in the Southern fleet, all 1968 Marshall-bodied buses acquired from Oldham. 5026 was repainted in 1973 after the divisional structure was in place and carries Selnec fleetnames. The ex-Oldham Panthers were not recertified when their original Certificates of Fitness expired and were withdrawn in 1974-75. The 422 service was the renumbered Oldham 22 to Moston. P Sykes

Consequent upon this change all new buses were registered in Manchester and were delivered with Selnec rather than divisional flashes. They were also allocated randomly across the divisions. The first block of registrations booked under the new system started with YNA271M which was allocated to 7271. The block YNA271-400M was actually booked for buses 7271 to 7400, reinstating the old-established practice of having fleet and registration numbers which actually matched. But that was not to be – at least not for a while – because both Selnec Northern and Selnec Southern had already booked numbers for buses which had already been allocated to them. So in the end the block YNA271-400M was allocated to 7271-9 and 7325-7445. Delays in deliveries meant that the last bus actually to be registered in this series was 7415 (YNA370M); subsequent deliveries had N-suffix registrations. The last four numbers in this series, YNA397-400M, were used on Leopard coaches 76-79, delivered in 1974.

Aside from this another block of registrations, XVU334-405M, was booked for single-deck deliveries. During 1973-74 it was allocated (but not fully used) as follows:

XVU334-366M Seddon midis 1704-1736
XVU367-386M Leyland Nationals 1300-1319
XVU387M Seddon-Chloride EX61
XVU388-393M Bristol LHs 1320-1325

The parcels operation was reformed as the Selnec Parcels Express Company, with a fleet of 108 vans, but experienced its first major downturn because of the loss of its biggest customer, Great Universal Stores. Profits in 1972 had been £19,182 but for 1973 fell to £2,381. From this point the parcels operation was on a downward slope.

Fares were raised in July – for the fifth time since Selnec was created – and fare scales were further rationalised and the number of different fare values reduced. The aim was to increase revenue by 11 per cent.

A standardised agreement covering all platform staff was reached with the trade unions in September. However the spread of one-man-operation, with the reduction in employment which came with it, did not always go smoothly and in October there was a three-day strike at Princess Road over omo conversions. The strike was triggered off by an omo conversion programme for the Flixton area which saw omo-equipped buses being transferred into Princess Road from Hyde Road, Northenden, Queens Road and Birchfields Road – and then being smartly transferred back out again. The problem was inadequate consultation with the staff.

At this stage Selnec was reconsidering its commitment to fare boxes on omo conversions. Many were still being done using buses equipped with fare boxes and on which tickets were not issued, but at the start of the year some existing omo services in Bolton were

Below **Northern's single-deck Fleetlines fared better than Southern's Panthers, giving up to 12 years service. 6090, one of three delivered to Bury in 1967, is seen in GMT ownership but still with Selnec fleetnames in Rawtenstall in 1975. Note the narrow exit door on the East Lancs body.** R Marshall

Below **Salford purchased its first two Daimler Fleetlines in 1963 and they operated for the PTE until 1975. The second of the pair, 4037, is seen in Eccles. The 66 from Peel Green to Manchester Piccadilly was an established Salford route which in 1972 was extended from Piccadilly to Ashton by joining it up with the 219, a route which in pre-PTE days was run jointly by Ashton and Manchester.** R L Wilson

Route 393 from Glossop to Mottram had been North Western's 123 and under Selnec control remained in the hands of ex-North Western buses. A Marshall-bodied Bristol RESL6G of 1968, one of 31 acquired by Selnec, heads for Mottram in 1973. R Marshall

Below **In the spring of 1973 six ex-Manchester Fleetlines were transferred from Selnec Central to Selnec Northern at Bury to replace ex-Manchester Atlanteans which did not sit comfortably in a predominantly Fleetline fleet. They were 1965 buses with Metro-Cammell bodies.** R Marshall

Below **By 1973 the number of buses in municipal liveries was declining rapidly. Former SHMD Fleetline 5644 with Walsall-style short front overhang was still green when photographed in Hyde bus station at the start of the year. The Northern Counties body was unusual in having a jack-knife entrance door and a sliding exit door. The ten buses of this type seated 68 and had Gardner 6LW engines.** R L Wilson

converted from fare boxes to ticket-issuing with motorised Setrights and 1973's omo conversion programme saw both systems being used. Bolton's rural omo services, incidentally, were always operated using ticket machines. The conversion of one urban route in the town from fareboxes to Setrights saw revenue rise by 10 per cent – confirming suspicions about fraudulent travel.

Vehicle transfers during 1973 saw further Rochdale Swifts, 6036/42-47, move westwards to Leigh, while Leigh Leopard 6064 was sent to Bolton. In the autumn three Bolton PD3s, 6622/30/32, went to Northenden while two Regent Vs, 6653/7, were moved to Oldham. The PD3s returned to Bolton in 1974. Bury received ex-Manchester 1965 Fleetlines 4701-6. These replaced six ex-Manchester 1965 37xx Atlanteans which had been with Selnec Northern since the spring of 1972. Two Bury Fleetlines, 6344/7, moved to Rochdale in February and stayed there until the autumn of 1976. Old municipal liveries were rapidly vanishing and all of the buses at Rochdale and Leigh were in Selnec orange and white by the summer.

Within Selnec Central there was a fair amount of movement between ex-Salford and ex-Manchester garages, most notably with forward-entrance ex-Salford PD2s making their way to Queens Road (16), Birchfields Road (12) and Hyde Road (one). In addition Salford Atlanteans 3129-34 went from Frederick Road to Queens Road in exchange for Mancunians 1042-48. More interesting arrivals in a Central garage were Rochdale Regents 6179 and 6207 which spent some time at Northenden in the autumn.

The prestigious Picc-Vic link, which had been given Parliamentary approval in August 1972, received a blow in August 1973 when transport minister John Peyton indicated that infrastructure grant for the work would not be available before 1975/76 because of the need to cut public expenditure. This was the start of a series of "on-off" announcements. In December Peyton offered government endorsement provided the Picc-Vic tunnel was part of a total transport strategy – which to most observers it quite obviously was. There were hopes that work would start in April 1975 with the line being open in 1980/81. Public expenditure cuts affected other PTE proposals including the upgrading of the Altrincham rail service and a variety of bus station projects.

Bus and train use increased slightly in 1973 with bus passenger journeys up from 523 to 524 million and rail journeys up from 22 to 23 million. The operation's 1972 surplus of £1.4 million was turned into a deficit of £1.4 million in 1973 as costs escalated. The number of staff employed fell slowly as one-man-operation was extended and stood at 10,413 at the year end, although in a buoyant employment market the staff turnover reached the unprecedented level of 24 per cent. The bus and coach fleet totalled 2,601 vehicles.

53

1974: GMT Takes Over

Local government reorganisation in 1974 saw the creation of Greater Manchester County Council. The role of the PTA – developing an efficient and integrated passenger transport system – passed to the new GMC. The Selnec PTE became the Greater Manchester PTE, responsible to GMC's transport committee, and the conurbation's buses were effectively returned to direct political control – or, depending on your viewpoint, interference. The PTE lost much of its independence as a result.

The change took place on 1st April and the new GMC area embraced Wigan, whose municipal bus operation was absorbed by the PTE on that date. This added 127 buses to the PTE fleet. Wigan's double-deckers were numbered in the series 3200-3339 which had originally been occupied by ex-Manchester Titans, the last of which had been withdrawn in 1971. Wigan's single-deckers were numbered between 1673 and 1691 at the end of a series (1600-1664) allocated, but not used, for a planned re-numbering of ex-North Western Bristol REs. Early Wigan repaints included Panther Cub 1676 and Leyland Titan PD3s 3219 and 3231.

In its four-and-a-half-year existence Selnec had taken delivery of 908 new buses and coaches and reduced the average age of the fleet from just under nine years to eight years. Of the 2,504 buses acquired from the original 11 municipal constituents 968 had been withdrawn, leaving 1,536 in service.

The oldest buses still in use when GMT took over were ex-Manchester NNB-registered Daimlers numbered from 4432 upwards. There were 17 and all were withdrawn in 1974; they were 20 years old. At the time of the GMT takeover there were still around 480 rear-entrance double-deckers in service, out of a total fleet of 2,753 vehicles. There were 12 chassis makes:

Leyland	1493
Daimler	946
AEC	125
Bristol	92
Leyland National	28
Seddon	25
Bedford	19
Metro-Scania	13
Dennis	8
Mercedes-Benz	2
Albion	1
Ford	1

The list of chassis makes was actually longer than it had been when the PTE was formed in 1969, but standardisation was making progress. Of the 908 buses delivered to Selnec a fair number were orders inherited from its constituents – but 384 (150 Atlanteans and 234 Fleetlines) were new 7000-series standard double-deckers.

The list of body makes was growing too – up to 22 from 19 in 1969. GMT took over buses and coaches bodied by Alexander, Burlingham, Crossley, Deansgate, Duple, Duple Midland, East Lancs, ECW, Leyland National, Longwell Green, Marshall, Massey, Metro-Cammell, Neepsend, Northern Counties, Park Royal, Pennine, Plaxton, Roe, Seddon, Weymann and Willowbrook.

Under GMC control the three Central divisions (East, North West and South) were retained but the Northern and Southern divisions were disbanded leaving only the district structure with Ashton/Stalybridge renamed Tameside and a new Wigan district. The new GMT districts to some extent reflected the district structure of GMC.

Between January and June two Bolton dual-door Atlanteans, 6789/90, were on loan to LUT. The lower deck seating capacity was reduced by two, making the buses 70 seaters (43 up, 27 down). They were fitted with Videmat ticket equipment and operated from LUT's Atherton depot on service 82, Bolton to Leigh. The Videmat system used a farebox into which passengers dropped their fares and a print of the coins appeared on the ticket as proof of the fare paid. In exchange two of LUT's dual-door Daimler Fleetlines, 399,400 (RTJ427/8L), were loaned to Selnec's Bolton depot.

Above **Unusual new additions to the Selnec fleet in 1973 were the first Mercedes-Benz buses to run in Britain. These were O.305s with Northern Counties 43-seat bodywork with sealed windows and forced-air ventilation. They initially ran for LUT but from mid-1974 were allocated to Oldham where 1355 is seen in GMT ownership.** J Robinson

Local government reorganisation saw Selnec become Greater Manchester Transport and absorb the all-Leyland Wigan municipal bus fleet. The first visible sign of the change in Wigan was a fleet renumbering with GMT fleet numbers being applied on an orange rectangle. 3218 was a 1960 Leyland Titan PD3 with 70-seat Northern Counties body. The services taken over from Wigan were renumbered in the 600 series. *M R Keeley*

Below An early Wigan repaint into GMT livery was 1676, one of two Massey-bodied Leyland Panther Cubs in the fleet. The only other Panther Cubs owned by GMT were two which had been in the Selnec Southern fleet and had come from Oldham and Ashton. 1676, devoid of fleetnames and still running with a Wigan route number, is seen in June 1974. *R L Wilson*

Left In the spring of 1974 two Bolton Atlanteans were loaned to LUT and fitted with Videmat ticket-issuing equipment. This registered the fare paid by producing a ticket which showed a facsimile of the coins which the passenger had dropped in the fare box. These stylish buses with fixed windows and sloping pillars were Bolton's only dual-door Atlanteans. They had East Lancs bodies. 6790 is seen in LUT's Atherton depot. *R L Wilson*

The loan to LUT of the two Mercedes-Benz O.305 single deckers, EX54/5, ended in May. They were then allocated to Oldham, where they spent the rest of their operational lives. They were initially used on route 159, Middleton to Woodhouses, which needed single-deckers to negotiate the low railway bridge at Middleton Junction.

The last fares increase under Selnec control was implemented in February and fare scales were further simplified in advance of the launch of off-bus ticket sales. GMPTE's first move in fares – in April – was to introduce a flat 2p concessionary fare on buses, to replace the previous half-fare bus concession, and a half-fare concession on trains. The flat bus fare was designed both to speed boarding and to give elderly people in Greater Manchester increased travel opportunities, both of which it achieved. The 2p fare also applied to children, replacing children's half fares. A second fares increase in 1974 was introduced in September. With it came further simplification of the fare stage structure.

In advance of the creation of the Greater Manchester PTE on 1st April experiments were carried out with the fleetname Manchester Metropolitan – reflecting the creation of the metropolitan county – which was applied to two Mancunian Fleetlines, 2189 and 2255, painted in a dark orange and white livery with different applications on each side of each bus. Neither ran in service. Other liveries considered were overall orange, and orange with yellow window surrounds. One PD2 (3465) was painted overall orange but, like the Fleetlines, was never let out into public view.

The PTE engaged a London-based industrial designer, Ken Hollick, to create a new identity to coincide with the change from Selnec to Greater Manchester and he produced the stylised 'M' logo which was designed to sit atop the word 'Metro' on a livery of overall orange with a white roof and brown wheels. Concern about the cost of a livery change – and a presupposition by some of the councillors sitting on the PTA that there would be no change – led to Ken Hollick's proposals being abandoned.

The Selnec livery layout continued unchanged but a slightly darker shade of orange – Metropolitan orange – was used to counteract problems with fading on the Sunglow which had been used by Selnec. The Mancunian white was replaced by a different shade. Hollick's brown wheels were adopted, as was the 'M' logo – but with the fleetname Greater Manchester Transport instead of Metro. The logo soon became known as the M-blem. The revised livery and M-blem appeared on new standards from Fleetline 7401, and was applied to existing buses when they were due for a routine repaint.

The fleetname and logo were applied on the lower side panels, except on the offside of rear-engined double-deckers where they were displayed on the panel which concealed the staircase. On half-cab double-deckers it was originally intended that the M-blem would be carried above the cab on the top deck side panels, with the Greater Manchester Transport fleetname in a single line under the front lower-deck side windows. This was applied to a small

In the spring of 1974 two Mancunians were repainted to represent four possible new liveries, with different schemes on the nearside and offside of each bus. Fleetline 2255 was painted overall orange. On the nearside it had two yellow stripes incorporating an "M" logo, located just above the lower polished moulding. On the offside the area between the lower moulding and the bottom of the lower deck windows was white, with two bands of brown relief and a similar "M" logo (below). The bus in the background, Fleetline 2189, was painted in a straightforward scheme with orange for the lower half and white for the upper half. The nearside had an orange roof; the offside a white roof. The two are seen in Hyde Road works. Neither was used in service in these trial liveries. *Alan Jervis/K Walker*

Left **On half-cab double-deckers it was planned that the M-blem and GMT fleetname would be displayed in a different style from that used on more modern buses. A large M-blem was to be carried between decks with the Greater Manchester Transport fleetname on the lower side panels. Only a handful of Titans received this layout before it was abandoned. These included 1964 ex-Manchester PD2 3718, seen in Stevenson Square in 1975. It was also applied to a few Stockport Titans.** *R Marshall*

number of ex-Manchester and ex-Stockport Titans, but in the end all buses were given the standard layout of fleetname adjacent to the M-blem.

At the end of April service revisions in Bolton saw the PTE take over Ribble routes to Dimple (B15, B25), Edgworth (206) and Walshaw (216). They became GMT 532/3, 563/4. In Bury GMT took over Ribble 255 to Edgworth which became 483. No vehicles changed hands but Bolton's single-deck allocation was increased with the addition of five new 13xx Leyland Nationals, quiet Metro-Scania EX60 and the five ex-Rochdale Fleetline SRG6LXs. LUT took over operation of GMT journeys on Bolton-based services 580/1 to Hulton Lane Ends, although these reverted to GMT in October 1975. The moves in Bolton and Bury were part of a complex series of route swapping which involved LUT, Ribble, Crosville and Warrington Borough Transport.

A new Centreline service started in Manchester on 1st July, linking Piccadilly and Victoria stations every five minutes (increased to every two-and-a-half minutes at peak times) and serving the main business and shopping areas of the city. A flat fare of 2p was charged with Almex tickets being issued. To maintain the fastest possible journey times the drivers were not allowed to give change.

To operate the Centreline – numbered 4 and running only from Monday to Friday – the PTE purchased more Seddon Pennine midis. The 22 delivered in 1974 (1715-36) had the revised style of Seddon body seen on 1704-14 but with only 19 seats, arranged in two-and-one layout to increase circulation space. The bigger side windows were of practical benefit to the 19 standees which could be carried. On the 1974 midis the livery layout was altered and featured a white skirt in place of the orange used on earlier buses, with orange for the window surrounds and roof. The mechanical specification was revised to include an Allison AT540 automatic gearbox, with the gear selector mounted on the driver's right as on the PTE's standards. These buses weighed 4569kg.

The small Perkins engine used in the Seddons was fairly noisy and 1734 was sent to the Sound Research Laboratories in Colchester to investigate methods of noise reduction. Operation of the Centreline was shared between Frederick Road, Queens Road, Hyde Road and Princess Road depots.

The Centreline – which was soon carrying over 30,000 people a week – was in many respects a substitute for the Picc-Vic line which was dead but wouldn't lie down. It appeared in GMC's first transport policy document, published in July, with the intention that work would commence in April 1975 as part of a £65 million rail improvement scheme. This envisaged electrification of the Manchester-Oldham-Rochdale line and re-electrification of the Bury line, converting it from third-rail to overhead electric operation.

October saw the introduction of dial-a-ride services in Sale. These were operated by Dial-A-Ride Ltd, formed jointly by the PTE and the coach-operating Godfrey Abbott Group. The operation used a fleet of six Bedford CF minibuses with 17-seat Deansgate bodies, CMA404-9N, and was described as the most extensive demand-responsive public transport system in the UK. In truth it had little competition – all of the dial-a-ride schemes being tried in the early 1970s were small scale operations. The Sale dial-a-ride services operated from 0700 to 1900, Monday to Saturday, and used a zonal charging system – the fare was 10p for travel in one zone or 20p (reduced to 15p off-peak) for two zones. Children were charged 5p but had to be accompanied by an adult. The service offered to pick passengers up within 30 minutes of a telephone call being made to the control centre. Sale was selected as an area where a high percentage of households would have a telephone.

The coach fleet, after a sizeable intake of new vehicles in 1973, received only four in 1974. All were 11m Leyland Leopards with Duple Dominant bodies. Of more significance was the transfer of coaches from Central and Southern division depots to the former North Western Road Car depot at Charles Street, Stockport, where coach administration was already centred. This took place in June and gave Charles Street an allocation of 17 coaches. Although Selnec had metamorphosed into GMT, the coaches continued to operate under the Selnec Travel banner.

The Manchester Centreline was an innovation which linked the city's two main rail termini by way of the central business and shopping districts. It started in July 1974 with a 2p fare – although by the time this photograph was taken in 1981 inflation had forced this up to 12p. A fleet of new Seddon midis with automatic gearboxes and an improved style of Seddon bodywork with bigger windows was bought for this service. 1732 arrives at Victoria Station. Stewart J Brown

Only four new coaches were purchased in 1974 and all were Leyland Leopards with Duple Dominant bodywork. R Marshall

New buses for the PTE included more double-deck standards. There were 90 Fleetlines which included (in May) the first standards to be allocated to Leigh. Atlantean deliveries re-started in August after an 18 month break and by the end of the year 25 had been taken into stock. All had 75-seat Northern Counties bodies. The original Selnec standard body had flat windscreens and problems were being experienced with the build up of dirt on the driver's rear-view mirrors. Wind-tunnel testing was carried out at the Motor Industry Research Association in Nuneaton jointly by Northern Counties and the PTE and as a result a new curved windscreen was adopted for what became known as the Mark 1A standard. The prototype was Fleetline 7348 and all new double-deck buses from 7401 featured the revised windscreen layout. Other features of 7348 included a Leyland National-style gear selector and spring parking brakes, both mounted on the driver's right, two-speed heater motors, carpet tiles, public address and the use of pvc in place of hide for the upper deck seat covering. The bus tipped the scales at 9348kg. The inclusion of public address and carpet tiles was partly with a view to running city tours – but that did not happen.

Other detail changes to the standard body came with 7365, on which the two rearmost hopper-type opening windows on the lower deck were replaced with top-sliders, a feature which appeared on subsequent standards. This was done to eliminate the risk of passengers sitting on the back-to-back seats over the rear wheelarches striking their heads on the protruding hopper frame.

With the standards came some other vehicles, most notably a batch of ten MCW Metropolitans delivered from October and allocated to Ashton. They entered service in December on the Trans-Lancs Express, replacing Leyland Leopards.

The Metropolitan was another joint venture between MCW and Scania – in effect a 73-seat double-deck version of the Metro-Scania with naturally-aspirated 202bhp 11-litre Scania D11CO6 engine, Scania automatic gearbox and – for the first time on a PTE double-decker – air suspension. GMT's ten were numbered 1425-34 to follow on from the renumbering planned for the Bristol VRTs delivered to Selnec Cheshire in 1973. The 14xx series was used for all subsequent non-standard double-deckers, following abandonment of the EX series.

Whatever the virtues of the ten Metropolitans no more were ordered. The combination of high weight (around 9820kg against 9350kg for a Fleetline) and a powerful engine took its toll on fuel economy with consumption in the region of 5.5mpg against 7.5mpg for a Fleetline. There was also the question of supporting local industry. The Metropolitan was an integral; Northern Counties could not body it. In addition, the Metropolitans cost around £21,000 each compared with around £17,000 for an Atlantean or £18,000 for a Fleetline with Northern Counties bodywork. The higher cost of the Fleetline reflected the price premium attached to Gardner engines which were generally held in higher regard than those available from Leyland.

Overall advertising continued to be a worthwhile source of revenue. 1974 Fleetline 7363 entered service in Stockport promoting the London and Manchester Assurance company. All of the 115 standards delivered in 1974 had Northern Counties bodies. This was one of the last buses with what became known as the Mark 1 standard body, identifiable by its flat windscreen. It is one of 350 Fleetlines ordered in 1971, delivery of which stretched out from 1972 to 1977. G R Mills

Ten Scania-powered MCW Metropolitan integrals joined the fleet in 1974 and were allocated to the Trans-Lancs Express. The livery layout was modified to suit the body mouldings, which led to the orange skirt being extended up to the lower-deck waist rail. The grille ahead of the rear axle was for the radiator. 1431 passes through Bury on its way from Bolton to Stockport. The square panel on the orange band above the nearside windscreen is an assualt alarm, activated by a button in the driving compartment. A sticker in the windscreen reads 'Alarm fitted'. M R Keeley

The Metropolitans wore a unique version of the Selnec/GMT livery. The lower orange area, usually restricted to the skirt panels, was extended up to the lower deck waistrail where there was a thick polished moulding as part of the standard body specification. One Metropolitan (1425) and one standard Atlantean (7511) were exhibited at the 1974 Commercial Motor Show. The Atlantean had an alloy frame (instead of steel), tinted windows and carpets. It was exhibited with registration BNE761N but was re-registered GNC292N before entering service.

The BNE-N batch of registrations was booked as the vehicle licensing system was being transferred to a new central computer at the Driver and Vehicle Licensing Centre in Swansea. It ran from BNE729-768N and was allocated as follows:

BNE729-730N	Seddon midis 1734-1735
BNE731N	Leyland National 1319
BNE732-750N	Standard Fleetlines 7416-7434
BNE751-758N	Standard Atlanteans 7501-7508
BNE759-768N	Metropolitans 1425-1434

However not all were actually used and vehicles which entered service from October were allocated registrations in the GNC-N series:

GNC276N	Seddon-Lucas midi EX62
GNC277-286N	Metropolitans 1425-1434
GNC287-289N	Standard Fleetlines 7422/23/27
GNC290-293N	Standard Atlanteans 7509-7512
GNC294N	Standard Fleetline 7428

During the confusion of the re-allocation of registrations 7423 and 7427 ran briefly with their intended BNE-N marks. The GNC-N series started at 275, which was a Leyland EA van in the service fleet. Seddon 1734 was delivered before August and received an M-suffix registration as originally intended. 1735 then received the registration BNE729N which had been earmarked for 1734.

A further Atlantean order was placed – for 176 – but largely because of continuing delays in the delivery of new double-deckers (208 were delivered in the 1974-75 financial year against a programme for 268) an order was also placed for 70 Leyland Nationals to be delivered in early 1975. The production line techniques used at the Leyland National factory in Workington were designed to cope quickly with large orders and, in any case, the factory was working well below its planned capacity.

These followed an initial batch of 20 Nationals, 1300-19, all 10.3m-long 41-seaters, and all but one delivered between January and March 1974. At this time there was uncertainty over the livery which would be used by GMT and the highly-automated Leyland National production line was unable (and Leyland's management unwilling) to cope with complex livery layouts: consequently 1300-19 were delivered in all-over white with the orange relief being applied at Hyde Road after the vehicles had entered service. The first 16 were delivered to Selnec; the last four to GMT. The 20th featured Leyland's new G2 automatic gearbox control and did not actually enter service until 1975. It was intended to have comprehensive engine encapsulation to create another Hush Bus – and the National needed it rather more desperately than did the Scania – but this did not happen. The Nationals were initially allocated to Oldham (nine), Altrincham (eight) and Bolton (three).

With the delivery of 20 production Nationals some of the original 1972 EX-series Nationals were redistributed with the first eight being moved from Bolton and Birchfields Road and being divided equally between Northenden and Weaste. At the same time the four Birchfields Road Metro-Scanias from the 1972 batch were moved to Northenden while Bolton's four Metro-Scanias went to Oldham and Ramsbottom (two each).

Six Bristol LH6Ls with Eastern Coach Works 43-seat bodies had been ordered by Wigan Corporation before its absorption by the PTE. These were delivered in GMT livery as 1320-25, following on from the Leyland Nationals. The LHs had Leyland's horizontal 125bhp 6.54-litre 401 engine, which was a development of the 400 engine already in the fleet in the ex-Oldham Tiger Cubs. The bodies were unusual in having a front-end with flat glass windscreens, a style which had been abandoned by ECW in 1970. Although delivered in May they did not enter service until January 1975 after much work was done on the suspension to improve ride quality. Whatever ride problems the LH buses had, it did not prevent GMT ordering two short-wheelbase LHS coaches for 1975 delivery.

A new numbering series for non-standard single-deckers was started in 1974 with the appearance of 20 Leyland Nationals as 1300-1319. Because no decision had been made on the livery to be used by GMT these were delivered in all-over white. They were 10.3m-long 41-seaters and were Selnec's first full-size single-door single-deckers. The 263 (previously 63) was an ex-Manchester service to Altrincham.
G R Mills

Wigan's last order for new buses specified six Bristol LHs with Eastern Coach Works bodies, a combination which would have been new to Wigan and was certainly new to GMT. The buses were delivered in May 1974 but GMT's engineers were unhappy with the ride quality of the LH chassis and they did not enter service until the start of 1975, after modifications had been made to the springs.
Stewart J Brown

Among vehicles ousted by the influx of new buses were the last of the ex-Manchester Panthers. Four were actually taken into GMT stock, but were sold by the end of the year. The small Panther Cub stock was depleted in 1974 as the surviving ex-Oldham and ex-Ashton examples were taken out of service. The Bristol RE, which was a contemporary of the Panther Cub, had proved an altogether more reliable bus and as E-registered Panther Cubs were being sold, two similarly-aged RESLs from the SHMD fleet, 5074/5, were transferred to the former North Western garage at Glossop for further service. The SHMD RESLs, of which there were six in all, were to outlast the Panther Cubs by five years.

Former North Western RESLs were on the move too, with four of the Marshall-bodied buses being moved into Weaste for operation on the 247 from Manchester Victoria to Eccles and Flixton. They stayed at Weaste for 12 months.

Dennis operation ceased during the year with the withdrawal of the two surviving North Western Lolines, 885/9, and the six which had come from Leigh, 6960-65. Two of the ex-Leigh buses saw brief service elsewhere in their last months, being moved to Rochdale in July and then to Princess Road in August and finally ending up for a spell at Altrincham. Three ex-Ashton Titans which were not long for this world – 1963 buses 5424/6/7 – went to Oldham at the end of the year and were withdrawn in 1975. Six Bolton Atlanteans, MWH-G-registered 6772-77, were transferred to Bury, while older buses from the same fleet, 1959 PD3s 6630-32 were sent to Stockport. 6630 soon moved on to Princess Road, its second stay at a Manchester garage – it had operated from Northenden for a period in 1973.

A few former Rochdale Regents were moved out of their home town in the latter part of the year with 6179 going to Altrincham and 6178/84 and 6204 going to Oldham where they were joined early in 1975 by 6216/7/21. 6178 also spent a short period at Hyde Road and 6179 at Princess Road. All were due for early withdrawal. Rochdale received four ex-Manchester 47xx Fleetlines which had been running in Bury for 12 months.

Top **The application of GMT fleetnames to Selnec buses was a gradual process. Ex-Manchester Titan 3695 still carries its Central logo and Manchester-style fleet number transfers in August 1974. It was one of 25 similar Metro-Cammell-bodied buses delivered in 1963 which were withdrawn by GMT in 1976.** G R Mills

Centre **In the early period of GMT's existence buses could be seen carrying either the GMT name or the Selnec lazy-S logo. In this view one of the ten short Fleetlines acquired from the SHMD has just been repainted and has the new names. The standard Fleetline behind still carries the Selnec name. Although designed as a one-man-operated bus, the short Fleetline was on a crew-operated service – note the 'Conductor-operated' sign alongside the fleet number.** M R Keeley

Right **Among the 22 different body manufacturers in the initial GMT fleet there was one bus bodied by Longwell Green: ex-Stockport PD2 5947 of 1960. It was withdrawn in 1974, but was then reinstated and survived until 1978.** G R Mills

60

Late in 1974 GMT introduced garage codes which were displayed on a white square alongside the fleet number on the front and rear of each bus. The codes were:

AM	Altrincham
BN	Bolton
BS	Birchfields Road (Manchester)
BY	Bury (and Ramsbottom)
FK	Frederick Road (Salford)
HE	Hyde Road (Manchester)
LH	Leigh
NN	Northenden (Manchester)
OM	Oldham
PS	Princess Road (Manchester)
QS	Queens Road (Manchester)
RE	Rochdale
ST	Stockport
TE	Tameside (Ashton, Stalybridge and Glossop)
WE	Weaste (Salford)
WN	Wigan

The garage codes which appeared on GMT's buses were rather more easily understood than the computer codes being used on internal documents. A series of three-digit numbers was used, 2xx for Northern Division, 3xx for Central and 4xx for Southern:

211	Bolton (Bridgeman Street)
212	Bolton (Crooke Street)
213	Bolton (Shiffnall Street)
221	Bury
223	Ramsbottom
231	Leigh
241	Rochdale
251	Wigan
311	Frederick Road
312	Weaste
313	Queens Road
321	Hyde Road
322	Birchfields Road
331	Northenden
332	Princess Road
333	Altrincham
411	Ashton
412	Stalybridge
413	Glossop
420	Oldham
440	Stockport

GMT was quick to link up with local radio to provide information on bus services and in November BBC Radio Manchester started broadcasting GMT service information five times a day.

Radios of a different kind were also of increasing interest. The entire Bury fleet was now equipped with two-way radios and plans were in hand to extend the system with control centres at Bury, Ashton and Devonshire Street, Manchester. The radio network would be divided into 17 zones – nine in the Central Division, three in Southern and four in Northern.

Staff turnover continued at a high level – 23 per cent – but shortages were reduced and services improved. Lost mileage was cut by around half. The change of structure from Selnec to Greater Manchester saw a change in accounting periods, moving from Selnec's calendar year, to the local authority financial year ending on 31st March. Consequently no financial results or passenger figures were published in 1974, with the first report of GMT and GMC not being produced until March 1975.

Top Selnec started life with six Dennises – all ex-Leigh Lolines in the Northern fleet. The oldest were two 1958 Mark I models with Gardner 6LW engines and East Lancs bodies. The six Lolines passed to GMT and were withdrawn in 1974. R L Wilson

Above There were still 16 ex-North Western Reliances in the fleet when GMT took over, all 36ft-long 1963 models with Willowbrook bodies. This one is seen in January 1975 in the former North Western depot at Altrincham; all had been withdrawn by the end of that year. R L Wilson

Left The initial GMT fleet had some 480 rear-entrance double-deckers, including this 1963 Leyland Titan PD2 with Roe body which had come from Ashton. R Marshall

1975: Rail Reversal

The 70 Leyland Nationals ordered in 1974 were delivered in 1975 and numbered 101-170, starting a new series for standard GMT single-deckers. All were, again, 10.3m 41-seaters. They had 160bhp 500-series engines and five-speed Pneumocyclic gearboxes with the new Leyland G2 automatic control system which had been on trial in National 1319. They left Workington in white and went to Northern Counties at Wigan where GMT's orange paint was added, along with other items such as GMT-style cab doors and two-way radios. Delivery started in March and finished in September and the buses were divided between Weaste (eight), Frederick Road (six), Bolton (five), Oldham (14), Stalybridge (four), Stockport (12), Altrincham (12), Birchfields Road (five) and Hyde Road (four). Stockport, Frederick Road and Hyde Road had not previously had allocations of Nationals.

A further six Seddon Pennine IV:236 minibuses entered service in April, bringing to 43 the number purchased since 1972. The 1975 deliveries, 1737-42, were similar to the previous year's with large windows, the simplified livery and Allison automatic gearboxes. These were the last Seddon midis built. GMT's vision of small buses penetrating residential areas and developing new services was ahead of its time. Interestingly, despite their small size and low weight, the intensive stop-start use facing the midis meant that their fuel consumption was only 7.5mpg, which wasn't very different from that of a full-size bus.

One very different Seddon midi arrived in 1975 – EX62, a battery-electric bus which was a joint venture between GMT, Seddon and Lucas. The Seddon body made extensive use of glass fibre panelling to reduce weight. EX62 had a 360V dc motor which developed 130bhp. But like all battery-powered buses it did have one serious drawback: even with glass fibre panelling it was heavy. EX62, a 19-seater, weighed 7738kg which was little less than some of the fleet's older double-deckers. By contrast the diesel-powered Seddon midis tipped the scales at only 4569kg. Before entering service on the Manchester Centreline in February, EX62 made history by being the first battery-electric vehicle to be driven non-stop between two British cities. It covered the 86 mile run from Birmingham to Manchester in just over three hours at an average speed approaching 30mph.

The other battery bus, Silent Rider EX61, entered peak-hour service in April on routes 202 and 203 between Manchester Victoria and Reddish. Manchester's interest in electric power did not end with battery buses and a report on trolleybuses was commissioned by GMC early in 1975. Nothing came of it.

Double-deck standards continued to arrive – 83 Atlanteans and 26 Fleetlines – all with Northern Counties bodies. Fleetline deliveries were being delayed by Leyland's large orders from London. Atlanteans from 7560 featured Leyland's new G2 automatic gearbox control and the model code changed from AN68 to AN68A. Wigan received its first two standards, Atlanteans 7562/3, in May. Included in the 1975 standards were registrations LNA251/2P (7598/99); coincidentally the PTE also operated LNA251/2G, which were Mancunians 2051/52.

Standard Atlantean 7071 was loaned to Hyndburn Borough Transport in March and 7580 was loaned to Blackpool Borough Transport in August. Similar bus 7592 was inspected by the Tyne & Wear PTE before being loaned to the South Yorkshire PTE for two weeks in October.

As a result of accident damage Mancunian 2246 and early standard 7012 were rebuilt with Mark 1A standard front ends with curved windscreens.

Most of the 150 surviving ex-North Western vehicles plus the 25 VRTs were scheduled to be renumbered at the start of 1975. Fleetlines were to be renumbered between 4051 and 4210, generally by adding 4000 to their existing fleet numbers. AEC Renowns were allocated 19xx numbers; Bristol REs 16xx numbers, and Bristol VRTs 14xx numbers. But the plan was soon modified. In the end some of the Fleetlines and all of the REs retained their old North Western numbers. The plan is outlined below.

Old No.	New No.	Type
400-424	1400-1424	VRT
270-275	1600-1605*	RESL
277-301	1607-1631†	RESL
315-344	1635-1664*	RELL
115-129	1915-1929	Renown
970-971	1970-1971†	Renown
1-6	4051-4056†	Fleetline
8	4058†	Fleetline
10-13	4060-4063*	Fleetline
15-21	4065-4071†	Fleetline
100-114	4100-4114	Fleetline
165-189	4165-4189	Fleetline
197-210	4197-4210	Fleetline

* 1603/4/37 and 4061 ran for a few months with their new numbers then reverted to their original numbers. Other vehicles in these batches were not renumbered at all.

† No renumbering was carried out.

Top left A very small number of buses reached GMT still in the livery of an original Selnec constituent. One was ex-Salford PD2 3008, seen in Warrington in July 1975. Salford eschewed external advertising and displayed the city's coat of arms on the upper deck side panels. Under Selnec ownership 3008's crest had been covered by advertising – but had re-emerged six years after the Salford City Transport fleet's demise when an advert had been removed but not replaced. 3008 was one of 38 Metro-Cammell-bodied Titans delivered to Salford in 1963. It was withdrawn, still in green, in 1975; the last buses of this batch survived until 1976. R L Wilson

Left Battery-powered Seddon-Lucas midi EX62 was delivered in 1975 and saw limited use. It was in theory renumbered 1362 but never actually carried its new number. It is seen in Doncaster in April 1977 on demonstration to the South Yorkshire PTE. R Marshall

Top The fifteen 1964 AEC Renowns acquired from North Western were renumbered in 1975 from 115-129 to 1915-1929, clearing the low numbers for re-allocation to new Leyland Nationals. 1915 loads in Manchester Piccadilly Gardens for the limited stop service to Glossop. R Marshall

Centre Also renumbered were the ex-North Western Fleetlines. All were allocated numbers in the 4000 series – recalling former Selnec Central Daimlers – but in the end the only vehicles to be renumbered were those whose numbers were likely to conflict with the new Leyland Nationals. 4209, formerly 209, was an Oldham-based bus and is seen in Stevenson Square, Manchester, on the service to Oldham and Waterhead. Like all the ex-North Western Fleetlines it has a lowheight Alexander body. M Fowler

Above Plans to renumber the 30 ex-North Western Bristol REs with Alexander Y-type bodies were aborted. New in 1970, these handsome 49-seat buses served GMT until 1981-82. On the service from Glossop, 335 heads for Padfield. The ex-North Western garage at Glossop was run as part of GMT's Tameside operation and its buses carried the TE depot code. R Marshall

Also renumbered in 1975 were the EX-series single-deckers. The four Seddons, EX56-59 had been renumbered as 1700-03 in 1974 (after 1703 ran briefly as 6059 in the Selnec Northern series). The remaining buses, EX30-55 and EX60-62, became 1330-55 and 1360-62 although the last two, both battery-electric buses, never carried their allotted 13xx-series numbers. To meet the demands of the PTE's computers the EX-series vehicles were latterly listed in the 90xx series but these numbers were never carried on the buses.

Significant transfers in 1975 included Bolton PD3 6629 which was moved to Hyde Road along with three ex-Wigan PD2A Titans, the first forward-entrance half-cab buses to run from the depot and the first ex-Wigan buses to be moved from their home town. Further Wigan PD2As found their way to Manchester garages during the year: two to Queens Road and three to Birchfields Road. The eight Wigan Titans in Manchester were withdrawn by the end of the year.

Former North Western buses on the move included 315/16 which introduced Bristol REs to Wigan; they were transferred from Altrincham in May – and returned there late in 1976. Three former North Western Fleetlines were moved to Rochdale from Altrincham and Princess Road. Rochdale also received two ex-Manchester 1963 VM-registered Fleetlines. Six ex-Salford PD2s were moved to Oldham, while four went to Stockport and three to Stalybridge. The last three replaced a trio of ex-SHMD Fleetlines which spent a short time at Leigh at the end of 1975, where they were not used before being moved on to Birchfields Road in March 1976. Quiet Metro-Scania 1360 (formerly EX60) was moved from Bolton to Oldham.

Withdrawals of note included the last Salford Daimler CVG6, although one CCG6 (with constant mesh gearbox instead of two-pedal control) soldiered on until 1977. The last ex-Bury half-cabs, five of the 1958-59 GEN-registered PD3 Titans, also left the fleet. The oldest ex-Manchester Fleetlines, 1962 and 1963 buses with reversed NE and VM registrations, were earmarked for withdrawal in 1975-76 and a number were, starting in May with 4607. However the slow delivery of new standards forced a rethink and ten withdrawn Fleetlines were recertified for a further five years service, most of which soldiered on until 1981, by which time they were a creditable 18 or 19 years old.

One batch of buses to disappear in 1975 after a remarkably long innings for a light-weight chassis was the group of four ex-Stockport Leyland Tiger Cubs which dated back to 1958. These had Park Royal-style Crossley bodies – and were the last Crossley-bodied buses in the fleet.

The first significant construction projects undertaken by GMT got under way in March – Altrincham Interchange and a new Tameside garage to accommodate 150 buses and replace existing ex-municipal premises at Ashton and Stalybridge.

In April a comprehensive programme of converting omo services from farebox operation, without tickets, to Almex ticket-issuing machines was started. This followed several years of farebox operation and was designed to cut fares evasion, produce more detailed statistical information and accelerate the omo conversion programme. Almost 70 per cent of the fleet was suitable for omo but only 55 per cent of the mileage was being so operated. The first of the Almex conversions took

GMT acquired 12 buses with side-gangway lowbridge bodywork – all East Lancs-bodied Titans from the erstwhile Leigh fleet. The last of the PD2s, new in 1958, were withdrawn in 1975. 6955 in Leigh bus station in 1975 shows the famous destination 'Dangerous Corner' and by some improvisation manages to show a three-digit route number in a box with a two-track blind display. A pair of ex-Leigh Renowns and an LUT Arab are visible in the background.
R Marshall

GMT's training fleet included a mixed bag of pensioned-off buses. TV7 was a Leyland Titan PD2 which had been new to Ribble in 1955 and had come to Selnec from North Western where it had been a training bus. It survived in the same role with GMT until 1980. Behind is TV2, a Longwell Green-bodied Titan which had been new to Stockport. The location is Charles Street depot.
R L Wilson

64

place in Oldham and Central division. By the end of the year all omo services at Altrincham, Birchfields Road, Frederick Road, Hyde Road, Northenden, Princess Road and Weaste were using Almex machines. The introduction of Almexes to Manchester and Salford followed their successful use on omo services in Stockport. The aim was for each driver to have his own Almex, although short-term delivery delays meant that this was not always achieved.

New double-deck models were being developed by leading chassis manufacturers in the mid-1970s. GMT already had ten MCW Metropolitans, delivered at the end of 1974. Contemporary with the Metropolitan was the front-engined Ailsa, built at Volvo's plant in Irvine in Scotland and powered by a small 6.7-litre turbocharged Volvo TD70 engine. The Ailsa had been developed at the behest of the Scottish Bus Group which, like most other operators, was dissatisfied with the performance of its rear-engined buses. The prototype, THS273M with Alexander body in the blue livery of SBG subsidiary Alexander (Midland), was used in service at Oldham in November.

In December NHG732P, a prototype of Leyland's advanced new B15 rear-engined integral double-decker, was inspected at Hyde Road but not used in service. At this stage in its development the B15 had a 170bhp Leyland 500-series turbocharged 8.2-litre engine, similar to that powering the National. It wore an all-over red livery.

A much smaller demonstrator was evaluated on the Manchester Centreline in August. This was a 27-seat Alexander S-type integral based on Ford A-series running gear. The vehicle tried alongside GMT's Seddons was GSA860N from the Grampian fleet. Elsewhere in the fleet's small bus business the Sale dial-a-ride operation was expanded in May to take in the entire town. However the Bolton Centreline service (503) ceased at the start of August, releasing Seddon midis which appeared on other routes in the town.

Against a background of concern about the availability of Leyland Atlanteans in sufficient numbers to satisfy GMT's needs a study of bus replacement policy was undertaken in the autumn. This identified the cost of a new Atlantean as £20,000, a Metropolitan as £23,000, and a Leyland National as £18,000. In each case half of the price was met by the government's 50 per cent grant towards the purchase of new buses. The Metropolitan was identified as the most expensive to operate; the National, rather surprisingly in view of the indifferent reputation of the fixed-head 500 engine, as the cheapest.

The report compared different vehicle types in the existing fleet and in an examination of fuel economy assumed 9mpg for front-engined double-deckers, 7.5mpg for rear-engined double-deckers (but only 6.5mpg for Metropolitans) and 9.1mpg for Leyland Nationals – a figure which seemed remarkably optimistic. A sample of vehicle availability at Hyde Road garage (195 rear-engined double-deckers, 60 front-engined double-deckers and seven single-deckers) confirmed engineering department claims that front-engined buses were more reliable. Where rear-engined double-deckers lost a frightening 35.3 per cent of their availability for service, front-engined models, even though they were generally much older, lost only 20.2 per cent.

Wigan's first Atlanteans were ten PDR1As with two-door Northern Counties bodies which were delivered in 1968-69 and most of which served GMT until 1982. M R Keeley

More typical of the Wigan fleet is 3240, a 1961 PD3A with a Northern Counties body, the design of which which was lengthened to fit the PD3 chassis by the insertion of a short centre bay – a procedure which a few other builders (most notably Park Royal with the London Routemaster) chose to follow. Alongside is a 1970 Leyland Panther, also with Northern Counties body. The PD3 was withdrawn in 1979 after 18 years service; the Panther in 1981 after 11 – which was more than most Panthers achieved. M R Keeley

No fewer than 18 new coaches were purchased in 1975. Two were Duple-bodied Bristol LHS6Ls (92/3) with the remainder being Leyland Leopards – 12 more with ECW bodies (80-91) and four with Duple Dominant bodies (94-97). Two of the Duple-bodied vehicles were 12m long PSU5 models; the other two were 11m-long PSU3s. As with previous deliveries they all had Pneumocyclic gearboxes. One of the 11m Duple-bodied Leopards had Stone-Platt air conditioning housed in a large pod on the roof at the rear. Not only was it the first fully air conditioned coach in the fleet – it was also one of the first in Britain. Among the vehicles displaced by the 1975 intake of coaches were the last of the three-axle Bedford VALs, 13/14, and the last Seddon coach, 47. The Seddon was reinstated for further service in the spring of 1976, finally leaving the fleet later that year.

The ECW bodies featured a route number display in the front dome; the previous batch had only had a destination display. They also used a much simpler – and rather plain – livery with white relief being restricted to the roof and corner pillars. They carried GMT fleetnames. By contrast the Duple-bodied coaches arrived with Selnec Travel fleetnames. Six of the ECW-bodied Leopards, 86-91, were loaned to LUT in October and used on stage services. They were taken out of service briefly after one was involved in a fatal accident, killing a conductress.

A few coaches ran for a short time with 00-series prefixes to their fleet numbers, but this practice was quickly dropped.

More orders were placed, this time for a further 80 Atlanteans and 120 Fleetlines. This would take Atlanteans up to 7935, while the Fleetlines – following on from an earlier order for 100 – were to be numbered from 8001 to 8220. In the end only 150 Fleetlines were added to the GMT fleet – 8001 to 8150.

Because Northern Counties was unable to meet all of the PTE's body requirements, an order for 160 bodies was placed with Park Royal. Although the intention was that all the Park Royal bodies would be on Atlantean chassis, continuing delivery delays of both chassis types meant that the PTE was prepared to have some of the Park Royal bodies mounted on Fleetline chassis if there were insufficient Atlanteans available. However that did not prove necessary.

Top In 1975 Bristols were introduced to the coach fleet – briefly trading as Selnec Travel – in the shape of two Leyland-engined LHSs with Duple Dominant bodies. One seated 29; the other 31.
P Sykes

Centre The Selnec Travel name was also applied to older coaches in the fleet, including this 1971 Bedford VAS1 with 29-seat Plaxton body in Charles Street depot, Stockport. Coaches carried the Charles Street address on the legal lettering, rather than the GMT headquarters address at Devonshire Street North. R L Wilson

Right The business of Warburton of Bury was taken over by GMT in November 1975. None of Warburton's vehicles were acquired but four coaches from the Selnec Travel fleet were given Warburtons Travel fleetnames. One of these was 1972 Leyland Leopard 52 with Duple Viceroy bodywork, seen on a private hire in Harwich.
G R Mills

The coach fleet had been renumbered in 1973 from the series 201-299 to 1-99. For administrative purposes the new series was 0001-0099 but it was not intended that the four-digit numbers would actually be carried on the vehicles. However a few did appear, including 0040, a Duple-bodied Bedford YRQ which entered service in 1972 and was among the last Bedfords purchased. It carries GMT rather than Selnec Travel fleetnames, indicative of demotion from the front-line coach fleet. The entrance has been rebuilt with a power-operated door, the front dome has been modified with a more generous – albeit still unused – destination box, and the fluted polished trim at skirt level has been removed. It is seen operating from Northenden depot on a peak-hour journey on route 103 from Manchester Piccadilly to Wythenshawe. R Marshall

In November the coach business of Warburton Bros (Bury) Ltd was acquired by GMT. None of Warburton's six coaches was purchased. Instead four existing Duple-bodied Leopards from the Selnec Travel fleet – 52, 67, 79 and 95 – were given Warburtons Travel fleetnames. Leopard 79 had reclining seats and acted as the Bury Football Club team coach.

At the same time GMT obtained an option to buy on or after 24th October 1980 an 80 per cent controlling interest in the Blundell group which operated 150 coaches through a variety of well-known subsidiaries, namely Blundell's Coaches (Southport) Ltd, Happiway-Spencers Ltd in Wigan, A E Hargreaves Ltd of Bolton, Smith's Tours (Wigan) Ltd and Webster Bros (Wigan) Ltd.

The Selnec Parcels Express operation was ailing. Its small 1973 profit – £2,381 – was turned into a substantial £95,509 loss for the 15 months to March 1975. This was attributed to inflation, recession and the fact that the 15 month accounting period embraced the business's slackest spells twice. Part of the response to this was to diversify and the company moved into van hire using a fleet of 20 Volkswagen, 20 Commer Walkthrough and six 4.5-ton box vans on Fiat and Commer chassis. Selnec Parcels Express also came to an agreement in May with TNT subsidiary Kwikasair to extend services to the Midlands and to London. The Parcels Express fleet stood at 120 vehicles operated from depots at Parrs Wood, Bennett Street and Haydock. Bennett Street depot was part of the GMT Hyde Road/Devonshire Street North site.

The Picc-Vic project had yet another uncertain year. In February the Transport Minister, now Fred Mulley, indicated on a visit to Manchester that all future transport plans should assume the line would be built. Thus encouraged, GMC produced a lavish brochure outlining what the line was and the benefits it would bring. The 2.75-mile long line would run from Ardwick Junction, just south of Piccadilly Station to Queens Road Junction, three-quarters of a mile north of Victoria. Just over two miles of the line would be in twin-bore tunnels with stations at Piccadilly Low Level (with an adjoining 12-stand bus station), Whitworth (under Whitworth Street), Central (under St Peter's Square), Royal Exchange (under the junction of Market Street and Cross Street) and Victoria Low Level.

Along with the £56 million tunnel (up from £26.9 million in 1971) there was £62.5 million-worth of other work. This included converting the Manchester to Bury line from 1200v DC third-rail electric to 25kv AC overhead, re-opening the line from Radcliffe to Bolton, electrifying two miles of the Buxton line between Hazel Grove and Stockport, and constructing a new depot at Bury. The Bury depot would carry out all maintenance on the 96 new three-car electric trains to be purchased. Four services were planned through the Picc-Vic tunnel:
• Bolton-Manchester-Styal-Wilmslow
• Bury-Manchester-Stockport-Alderley Edge
• Victoria Low Level-Stockport-Hazel Grove
• Victoria Low Level-Stockport-Macclesfield.
These would run every 10 minutes at the peak and every 30 minutes off-peak, giving a peak frequency through the tunnel of a train every 2½ minutes.

GMC's brochure on the line concluded lyrically: "The Picc-Vic project is not just a tunnel, nor just a railway network. It is the first part, probably half, of a complete package of improved public transport which will offer to the travelling public for the next 100 years a system that is a realistic alternative to the car. It would seem a tragedy for the project to founder because the total impact on the urban fabric of Greater Manchester is either misunderstood or misinterpreted."

Tragedy was in fact just around the corner. Later in the year the Transport Minister – now John Gilbert – refused to authorise the project.

However it was agreed that the Manchester to Bury line be converted from third-rail to overhead electric operation. The work was to start in 1976 with new trains operating from 1979, replacing the Class 504 two-car sets which had been introduced to the line in 1959. The work on the Bury line would be followed, starting in 1977, with the electrification of the Manchester–Oldham–Rochdale service which had earlier been listed for closure. The plan was to introduce electric trains in 1980. It didn't happen.

Fares rose twice in 1975 – in February by around 28 per cent and again in August by a swingeing 45 per cent. The August rise saw the end of fareboxes, the abolition of cheap weekday fares and a doubling of the concessionary fare from 2p to 4p. The end result was a not unsurprising 20 per cent drop in ridership. The replacement of fareboxes by ticket-issuing equipment exacerbated the effects of the fares increase insofar as a significant minority of bus users had been underpaying for their journeys before the advent of tickets made it possible for inspectors to check on over-riding.

To help combat the passenger losses a new SaverSeven weekly ticket was introduced in October. This allowed unlimited bus travel throughout Greater Manchester for £2.50 a week and was an instant success with 35,000 sales a week within six months of its launch. By this time 15 per cent of adult passenger revenue was coming from SaverSeven tickets. Research showed that one-third of SaverSeven ticket buyers were making more use of buses and that a small number – four per cent – claimed to have been mainly car users before switching to SaverSeven.

The centralisation of vehicle licensing at Swansea had seen the Manchester local vehicle licensing office issue blocks of numbers with considerable parsimony and during 1975-76 new buses were registered in small batches. This changed in November and from Fleetline 7470 and Atlantean 7600 much larger blocks of numbers were made available to GMT – and they were at long last booked to match the bus fleet numbers.

Changes in GMT's management saw Jim Batty, who had assumed responsibility for all of the outer district operations, move to the post of chief operations executive. His place was taken by Tom Dunstan. Ken Holt relinquished control of service operations and became responsible for the impending integration of LUT and GMT. Revised accounting procedures saw a 15 month financial period to 31st March 1975 during which 677 million bus and 39 million rail journeys were made. The bus fleet at the end of the period stood at 2,804, swollen by the absorption of Wigan. GMT employees numbered 11,502 and staff turnover was still high at 23 per cent.

1976: LUT Takeover

On 1st January the PTE exercised its option to take over Lancashire United Transport, Britain's biggest independent bus and coach operator, for £2.65 million. LUT ran 363 buses and coaches from depots at Howe Bridge in Atherton (which also housed the company's head office and works), Swinton and Hindley.

The fleet was varied and a conservative buying policy in the 1960s meant that the majority of its double-deckers were half-cab Guy Arabs, of which there were 140, the newest dating from 1967. The later vehicles had forward-entrance bodies. LUT's conservatism did not extend to total rejection of rear-engined buses and the oldest of the 55 Daimler Fleetlines acquired by the PTE dated back to 1962. LUT's newer Fleetlines were 10m-long two-door models. Ironically LUT had started buying 10m-long two-door 'deckers at around the same time as Selnec concluded that the 9.5m single-door type was the bus of the future.

The single-deck fleet was somewhat quirky. The oldest were a couple of AEC Reliances; the newest, Bristol REs. In between LUT had bought Leopards and Tiger Cubs (as late as 1967 and with dual-door Marshall bodies for the last of the Tiger Cubs) and Seddon Pennine RUs of which there were no fewer than 50, all with Gardner engines and two-door Plaxton bus bodies. The oddest of its single-deckers were 20 mid-engined Bristol LHLs, also fitted with two-door Plaxton bodies.

Under PTE ownership LUT continued to run with a high degree of independence. The fleet retained its red and grey livery and original fleet numbers, even though many clashed with numbers allocated to GMT buses.

The coaching unit had retained the Selnec Travel name after the Selnec PTE ceased to be. In March 1976 a new name was coined – Charterplan – and with it came a new livery of white with orange and brown stripes which were upswept at the rear in a style which echoed the car used by the stars in the American TV police series Starsky and Hutch, but was in fact copied from a French coach operator. The stylish upswept stripe was soon imitated – but never with quite the same panache – by coach operators up and down the country. Leopard 96, new in 1975, was the first coach to receive Charterplan colours. It had previously been in a white, red and green promotional livery for tailor-made package tours.

The Charterplan name ousted Selnec Travel, but the Warburton Travel name was retained on blue-striped coaches. LUT coaches were painted in a variant of the livery which used red and yellow stripes. The dark brown used in the Charterplan livery was also to find an application on the bus fleet where it replaced the mid-brown used on bus wheels.

The SaverSeven weekly bus tickets launched in November 1975 were extended from April to include rail travel. The original £2.50 ticket now allowed rail travel within a central zone while tickets at £3 and £3.50 allowed more extensive rail use. A £4 SaverSeven covered all rail and bus services in the county. To help speed boarding by people using pre-purchased tickets GMT ordered 3,000 Almex M ticket cancellors.

The PTE's original director general, Tony Harrison, left in July to become chief executive of Greater Manchester County, but remained a non-executive director of the PTE. He was succeeded in the top PTE post by David Graham, who had been director of finance since the formation of Selnec.

The Picc-Vic tunnel finally ground to a halt in 1976 when it became clear that the necessary government funding was never going to be made available and the project team was disbanded – this after expenditure of £3.1 million over the previous few years. The only positive rail-associated activity in 1976 was the introduction in August of a rail feeder service, the 510 Leverlink, running on weekday peak periods from Little Lever to the stations at Radcliffe and Farnworth.

Above **The last buses delivered to LUT before the takeover by GMT were five Plaxton bus-bodied Leopards. Two were 44-seat dual-door PSU3s and the other three were single-door 44-seat PSU4s. All were of an improved design with double-curvature BET-style windscreens, bigger side windows and a larger area of grey relief.**
R L Wilson

Left The Lancashire United fleet included some strange buses. Seen in Leigh bus station in 1972 on the Horwich service is 155, one of eight Plaxton-bodied Leyland Leopards delivered in 1964 and among the oldest single-deckers in the fleet when GMT took control. **Right** LUT was among Seddon's biggest bus customers with a total of 50 rear-engined RU models, all fitted with 40-seat Plaxton bodies. They entered service in 1970-71 and 365 is seen in Wigan.
G R Mills/Stewart J Brown

Left A one-off in the LUT fleet was this 1974 Leopard with Northern Counties 44-seat bodywork with fixed windows and an unusually wide entrance for a Leopard bus of this period. It was an exhibit at the 1974 Commercial Motor Show and is seen leaving Leigh for Wigan in the summer of 1975. **Right** Most of LUT's single-deckers had Plaxton Derwent bodies. Ten delivered in 1974-75 were based on Bristol RESL6G chassis and had 41 high-backed seats. R L Wilson

Left The new-look Charterplan livery, to replace Selnec Travel, was introduced in 1976. Air-conditioned Leopard 94 in the new colours – white, orange and brown – drives off the Scrabster ferry on Scotland's north coast in 1978 on a private hire to the Faroe Islands. The Duple Dominant body is unusual in having flat glass double-glazed side windows. **Right** 1974 Duple-bodied Leopard 76 in the Warburtons fleet received a two-tone blue version of the Charterplan livery. The blue stripes are painted over the polished waist moulding. R L Wilson/R Marshall

Deliveries of new standards continued to run behind schedule: 7500, the last of the 1973-74 Fleetlines, entered service at Rochdale in December, two years late; 7679, the last of the 1974/75 Atlanteans was delivered to Wigan in October, 12 months late. In all 108 standards were delivered in 1976 – 81 Atlanteans and 27 Fleetlines, all bodied by Northern Counties. Glossop got its first standard, a Fleetline, in April. In an attempt to ensure continuity in the supply of new buses an order for 400 Atlanteans was placed with Leyland for delivery between 1978 and 1980.

Standard GMT buses were still of interest to other operators and Atlantean 7606 was inspected by Brighton Borough Transport in March while similar bus 7628 spent a few days with the West Yorkshire PTE in May. Seddon Pennine battery-electric midibus EX61 was loaned to the South Yorkshire PTE in April and used in service in Doncaster.

Two further experimental buses were delivered in 1976 – the first double-deck buses to be built by Foden for over 20 years. The Foden-NC, a joint venture between the Sandbach truck builder and Northern Counties, was a belated attempt to produce a bus to combat Leyland's monopoly – belated in that the Ailsa and the MCW Metropolitan had been launched two years earlier.

The Foden-NC had a welded perimeter frame with a 16ft 9in wheelbase. It was powered by a transverse rear Gardner 6LXB engine which was coupled to an Allison MT640 torque convertor automatic gearbox, a rare application of an Allison gearbox in a double-deck bus. A Ferodo retarder was fitted, making the Fodens the first buses in the GMT fleet to be equipped with an item that was soon to become a standard feature.

The Northern Counties bodies were basically adaptations of the Selnec/GMT standard with 75 seats in the usual 43/32 split. A Foden badge adorned the front panel while the rear end, with valances between the engine compartment and the upper saloon floor, echoed a style developed by Northern Counties almost 15 years earlier to give a smoother rear profile to rear-engined buses.

The Fodens were added to the 14xx series of non-standard double-deckers as 1435/36. The first was received in June; the second was exhibited on the Northern Counties stand at the 1976 Commercial Motor Show, the last to be held in London's Earls Court. Neither entered service until 1977. Also on show at Earls Court, in the demonstration park, was a new Duple-bodied Bristol LHS6L coach, 98, the last Bristol purchased by GMT.

In connection with service changes and omo extensions in Leigh, all 13 Metro-Scanias were moved there in April from Stalybridge, Oldham and Bury. Allocating the entire Metro-Scania fleet to one depot simplified maintenance and spare parts stocks. These, and Leigh's other single-deckers – Rochdale-style Swifts 6040-49 and ex-Leigh Corporation Tiger Cubs 6060-63 – had their capacity reduced by one seat (on dual-door buses) or two seats (on the Tiger Cubs) to make space for ticket cancelling machines.

Leigh had earlier reduced the seating capacity on its five 1964 AEC Renowns (randomly numbered 6908/30/34/38/39) by removing the front upper deck seat. This cut the seating from 72 (41 up, 31 down) to 69 (38/31). These buses were then transferred to Wigan, giving the town its first taste of AEC operation.

Other transfers of note included three SHMD NMA-D-registered Fleetlines which moved from Tameside to Manchester, with one at Northenden and two at Princess Road. Four ex-Bolton ABN-C-registered Atlanteans were moved to Manchester and nine dual-door buses from the OBN-H batch with sloping pillars were transferred to Wigan which had standardised on dual-door buses in the period prior to its takeover by GMT. Five of the ex-North Western Fleetlines with distinctive Alexander lowheight bodies were transferred to Birchfields Road from Oldham in the spring. At the same time RELL 320 moved from Altrincham to Glossop.

Services in Marple were reorganised in September and omo was extended further. To achieve this six standard Atlanteans were moved into Stockport to replace six of the ex-North Western Fleetlines which were re-allocated to Birchfields Road. At the same time ex-Stockport Corporation forward-entrance Titan 5893 moved to Northenden while 5894 went to Bury.

The new Altrincham Interchange was opened in November, and services were revised at the same time. To coincide with this, and the launch of new Interlink services, Altrincham received an influx of standards in the shape of 12 Fleetlines. These replaced 13 ex-North Western Alexander-bodied Fleetlines of which one each were dispersed to Princess Road, Northenden, Birchfields Road, Hyde Road, Queens Road, Frederick Road, Weaste, Ashton, Stalybridge, Oldham and Rochdale while two went to Bolton.

The Interlink services were designed where possible to connect with trains to and from Manchester and some were increased in frequency from every 20 minutes to every 15 minutes to co-ordinate with the train timetable. The services were operated by Nationals and double-deck standards with black-on-yellow destination displays.

Bus fares were increased in October with a minimal loss of passengers – around one per cent, compared with 20 per cent following 1975's two rises. There was no room for complacency as passenger figures were continuing to fall and as part of its drive to improve services GMT started to monitor travel patterns. This was done by regular sampling of passengers and the introduction of household travel surveys – and against a background of a service network which was largely unaltered from municipal days.

To obtain full control of the Sale dial-a-ride operation GMT acquired the Godfrey Abbott Group on 1st November. Godfrey Abbott ran 26 coaches (mostly Bedfords), six rather dubious second-hand double-deckers and 10 Bedford CF minibuses. It also ran a licensed Manchester to Paris coach service, making GMT the first PTE with an international service. A new variant of the Charterplan livery layout, with two-tone green stripes, was used for Godfrey Abbott coaches. The Godfrey Abbott Group included four non-trading subsidiaries – Lingley's Sale-Away Touring Co, Pride of Sale Motor Tours, Stretford Motors and one-time North Western subsidiary Altrincham Coachways.

The first of the two Foden-NCs operated by GMT was completed in the summer of 1976 and is seen here at Hyde Road garage in July. It did not enter service until February 1977. The Foden-NCs carried a badge on the front, unlike most other standards, and had the rear engine enclosed within the bodywork. They were slightly longer than GMT's standard Fleetlines and Atlanteans and the extra length was accommodated in a longer rear top deck side window. 1435 lacked the detachable shallow skirt panels which were a feature of the normal standard body. J Robinson

Right **The Godfrey Abbott fleet wasn't all glamorous coaches. 661KTJ was an early PDR1 Atlantean with lowbridge bodywork by Weymann. It started life in 1959 as a Leyland demonstrator and reached Godfrey Abbott in 1972 by way of Bamber Bridge Motor Services and Ribble. It was sold at the end of 1977.** R L Wilson

Far right **Godfrey Abbott's other double-deckers included two East Lancs-bodied Albion Lowlanders which had been new to Luton Corporation. One survived as a store until 1980.** R L Wilson

The Sale dial-a-ride service was started as a joint venture by GMT and the coach-operating Godfrey Abbott Group. The first generation of minibuses used on the service were ten Bedford CFs with angular coachbuilt Deansgate 17-seat bodies. JJA173N is seen in Sale in August 1975. In November 1976 GMT took control of the Godfrey Abbott Group. R L Wilson

Grand plans were announced to launch a dial-a-ride service in Altrincham in 1977 which would have needed up to 50 small buses – but the service never materialised.

Repainting the fleet was a lengthy process and it was only at the end of the year that the last ex-Wigan Atlantean received GMT colours. But at the same time 1961 Titan 3230 was still running in Wigan livery displaying its old fleet number, 57.

Selnec's expectations that services could be developed in town centres and quiet suburbs which would be operated by Seddon midibuses had proved a trifle optimistic and at the start of the year no fewer than 13 were taken out of service as surplus to requirements. These included 1700-3, the original 1972 prototypes. Two other small buses were withdrawn during the year: the ex-Bury Bedford J2 6081 and the ex-Ramsbottom Albion Nimbus 6082. The latter had been the PTE's only Albion.

Fully-fronted double-deckers disappeared from the fleet in 1976. Selnec had inherited 27 Leyland Titans with full-front bodies from Bolton. The first ten, new in 1961, had been withdrawn in 1973. But the remaining 17, 1963 PD3A/2s, passed to GMT and came out of service during the year. They were also the last ex-Bolton front-engined buses and their withdrawal was immediately followed by that of the oldest ex-Bolton Atlanteans, also new in 1963.

During 1976 Leyland was talking to a number of major operators about its new B15 integral double-decker. GMT had inspected a prototype in 1975 and announced at the end of 1976 that it was taking an option on 50.

The year ended with a bad vehicle shortage at LUT's Swinton depot as the 50 Seddon Pennine RUs delivered in 1970-71 started to come up for recertification. This necessitated the temporary transfer to Swinton of an assortment of distinctly unusual vehicles for LUT. These comprised a Wigan PD3 (still in Wigan livery), five Manchester open-platform PD2s new in 1958-59 along with four 1963-64 Stockport PD2s and two Leigh lowbridge Titans. The loaned vehicles ran for LUT until the middle of February 1977.

A more permanent sign of GMT's ownership of LUT came at the end of the year with the start of a programme to replace TIM ticket-issuing machines with the standard GMT Almex.

There was also a vehicle shortage looming in the main GMT fleet, thanks to the slow delivery of new standards. Accordingly October and November saw around 25 time-expired buses being recertified for further service. These included 1962 Manchester Fleetlines, 1963 Stockport PD2As, 1963 Manchester CVG6s, 1964 Salford PD2s, 1958/59 Manchester PD2s and 1963 Bury Atlanteans.

The parcels operation continued to lose money – £73,908 in 1975/76, while the PTE's bus services continued to lose passengers. In 1975/76 they carried 483 million passengers while the bus fleet, swollen with the absorption of LUT, crossed the 3,000 mark for the first time to a peak of 3,065. The number of employees peaked too, at 12,500. From this point on, both vehicle and employee numbers would fall.

1977: New Deliveries Pick Up

In 1975, while still independent, LUT had placed orders for new double-deck and single-deck buses for delivery in 1976-77. These called for 15 Leyland National 11.3m buses (LUT's first), 30 Leyland Leopard PSU3s with Plaxton Derwent bus bodies and 25 Bristol VRTs with 9.5m-long Northern Counties bodies.

The Plaxton-bodied Leopard buses, PSU3s with Pneumocyclic gearboxes and 48 seats, were delivered in 1976 as 435-64. They were followed by the Nationals which arrived in 1977 as 465-79 and were delivered in all-over London Transport red. They had fully-automatic Pneumocyclic gearboxes.

The VRTs would have been a new type for LUT and were ordered largely because of delays in getting Daimler Fleetlines: the VRT was seen as the next best thing. GMT cancelled the Bristol order and in its place substituted an order for 45 Fleetlines with Northern Counties bodies to GMT standard specification. Delivery of these buses, in LUT's red and grey livery, started in 1977 and continued through to 1979. A plan to number them 8001-45 in a new GMT series was not proceeded with and they were delivered as 485-529 with three distinct registration groups – OBN502-11R (delivered in 1977), PTD639-58S (1978) and TWH690-704T (1978-79).

LUT's new 1977 coaches, Leopards 480-84, were delivered in the new white coach livery with upswept red and yellow relief bands applied in the Charterplan style. They retained Lancashire United fleetnames, but with the GMT M-blem alongside. A further sign of the integration of LUT and GMT was the adoption of orange LUT publicity material featuring the M-blem. There were no new coaches for Charterplan in 1977, but Godfrey Abbott got three AEC Reliances. These had Duple Dominant bodies with 44 reclining seats and a toilet compartment and were allocated to the Manchester to Paris service which was now operating five times a week.

Above **LUT's first Leyland Nationals, ordered before the takeover by GMT, were delivered in all-over red in early 1977. There were 15, all 11.3m-long 49-seaters. Two further batches of 15 similar buses were delivered in 1978 and 1979, bringing to 45 LUT's National fleet.** R L Wilson

Below **An LUT order for Bristol VRTs was cancelled by GMT and in its place LUT received new Fleetlines based on the standard GMT specification but in LUT livery and with LUT interior trim.** R L Wilson

GMT was one of a number of major operators to mark the Queen's silver jubilee in 1977 by operating a few silver-painted buses which were sponsored by advertisers – the Daily Mirror in the case of standard Atlantean 7706. It was one of six silver jubilee buses – all new Atlanteans – operated by GMT, who also painted silver buses for other operators. R L Wilson

In the main GMT fleet new standards were still being delivered. There were 173, the highest number since 1973. Northern Counties supplied 109, but from June Park Royal, whose last deliveries had been in 1973, reappeared with the first of an order for 160 bodies on Atlantean chassis which started at fleet number 7801. By the end of the year Park Royal had delivered 64 buses. All of the 1977 standards were Atlanteans.

A further 15 Nationals, 171-185, were ordered for the GMT fleet to alleviate the effects of late deliveries of double-deckers. These broke new ground in that they were long 11.3m models rather than the PTE's standard 10.3m version. They were taken to speed delivery. They were delivered in March and were allocated to Rochdale (five), Oldham (five), Altrincham (two), Birchfields Road (two) and Bolton (one). They were again delivered from Workington in white and painted by the PTE. Recertification of buses due for withdrawal continued as another way of counteracting the shortfall in new bus deliveries.

It was becoming clear that the days of the Fleetline and Atlantean were numbered. Just as the integral Leyland National had swept away other Leyland group rear-engined single-deck chassis, so the new B15 was planned to replace existing double-deck models. GMT was still committed to the old models and had in 1976 announced an order for a further 400 Northern Counties-bodied Leyland Atlanteans. But at the same time it was examining the other options open to it.

The two Foden-NCs entered service on 4th February from Princess Road depot and represented one option for the future. Another option, on 50 of Leyland's new B15 double-deckers, was confirmed as an order, and an order was placed with MCW for ten of its new Metrobus semi-integral double-deckers. The Metrobuses were to have Gardner 6LXB engines with Voith gearboxes in five and GKN-SRM gearboxes in the other five.

A smaller order in all senses of the word specified eight Bedford CF340s and two Ford Transits with Reeve Burgess 17-seat bodies to the angular Reebur style. These were for the dial-a-ride operation in Sale. This was extended in October to include evening services, marketed as Nightrider, which ran from 1900 to 2400.

Delivery of new standards continued apace. The last few of the original order for 350 Daimler Fleetlines arrived at the start of 1977 and with them came the last 120 of a batch of 300 Northern Counties-bodied Atlanteans and the first of 160 Park Royal-bodied Atlanteans. Hyde Road's 7776 represents the Atlantean/Northern Counties combination and is seen amidst dereliction at Victoria Station before the area was landscaped and the station frontage cleaned. An ECW-bodied Leopard on the airport express stands behind. M R Keeley

Small numbers of vehicles were still being transferred to unfamiliar towns. Three ex-Oldham GBU-D-registered Atlanteans went to Bolton for a short spell, while four of Bolton's own Atlanteans returned from a period in Manchester. Three of Tameside's ex-SHMD Fleetlines from the GTU-C batch were moved to Rochdale, and one similar bus spent a short time in Bury. Two ex-Leigh Renowns were transferred to Manchester (Birchfields Road).

Former Manchester Fleetline 4746 moved from Northenden to Bury where it was soon joined by nine of the single-doorway East Lancs-bodied Mancunians from Queens Road. Bury released ex-Stockport Titan 5894, which returned to its home town. Other ex-Stockport Titans being moved during the year included two 1963 PD2As from Oldham to Weaste and then on to Queens Road at the end of the year, and a pair of 1967 exposed-radiator PD2s which were drafted in to Oldham. Four of the PD2As moved from Princess Road to Stalybridge. Ten of Salford's dual-door Atlanteans of 1969 vintage were transferred to Wigan which, after Manchester, had GMT's biggest fleet of dual-door buses. North Western Bristol RELLs on the move included two to Princess Road and one each to Stalybridge and Rochdale.

A few buses still survived in municipal liveries. Manchester Fleetlines 4600/8, new in 1962, were withdrawn in 1977 still in Manchester's red livery – almost eight years after the formation of the PTE, although they had in fact been repainted red during that time. At Oldham, one 1957 Roe-bodied PD2 was still in Oldham Corporation colours, but had been in the training fleet since 1974. Wigan Titan 3230 retained its original livery and fleet number, 57, only being renumbered (using Wigan-style shaded numerals) when Leopard coach 57 was transferred to the depot.

Withdrawals in 1977 included 12 of the first 15 production Seddon midibuses (which had manual gearboxes), leaving only the Centreline fleet plus three at Ramsbottom. These were comparatively modern buses (dating from 1972-73) and soon found new homes. Six were sold to Tayside Regional Council but two found what was a rather ignominious new role for such young buses as staff transport for a builder.

The last of only 25 side-gangway lowbridge buses to have been owned by Selnec and GMT, Leigh's 6937, a 1962 East Lancs-bodied Titan, was withdrawn in July. The last ex-Oldham half-cabs, nine handsome Roe-bodied PD3 Titans, were withdrawn during the year, as were the last SHMD Daimler CVG6s. At Rochdale the last of the ex-municipal Weymann-bodied AEC Regent Vs, a type characteristic of the town for two decades, were withdrawn in July. The last North Western Renowns went in October – eight had survived into 1977, all at Glossop – and this left the 11 remaining Leigh Renowns as the only AEC double-deckers in the fleet. The chassis of Mancunian 2225 was allocated to the apprentice training school after its body was destroyed by fire.

The Seddon-Chloride Silent Rider wasn't exactly withdrawn, but it wasn't exactly in regular service either; neither was the Seddon-Lucas electric midi. During 1975-76 the two buses achieved availability for service of only 50 per cent, compared with 80 per cent for diesel buses. Of 445 days required in service in 1975-76 the Silent Rider was in service for 84, on demonstration for 80 and failed for 281. The Seddon-Lucas bus did better with 173 days in service out of a target 475. It spent 28 days on demonstration and failed on 274 days. One of Silent Rider's rare forays out in 1977 took it rather further afield than Reddish, which had been its usual haunt. It was shipped to the USA where it was an exhibit at the 1st International Electric Vehicle Exposition in Chicago. It is perhaps unkind to ask why GMT bothered to ship it back again – but it saw no further use in service.

The long-established link between SHMD and front-engined Daimlers was broken in 1977 with the withdrawal of the last survivors still running for GMT. These were six 1964 CVG6s with 64-seat Northern Counties bodies. R L Wilson

However the SHMD-Daimler link was maintained for a few more years by Fleetlines. 5635, seen in Piccadilly Gardens in 1977, survived until 1980 and the last of the ex-SHMD buses did not go until 1981. New in 1966, 5635 had a Northern Counties body with valances above the engine compartment to give the appearance, from the side at least, of a smooth profile to the rear end. M R Keeley

A few ex-Bolton Atlanteans spent a spell at Manchester's Princess Road garage in 1977, including 1965 Metro-Cammell-bodied buses 6720-22. 6720 is seen in Piccadilly in August; it and the others were back in Bolton by the end of the year. M Fowler

The 25 ECW-bodied Bristol VRTs which had been ordered by North Western spent their lives at Stockport and added variety to the modern bus scene in Manchester, which was increasingly being dominated by PTE standards. M R Keeley

In May there were extensive revisions to rail timetables with a regular headway service on the Manchester Victoria–Bolton–Atherton–Wigan line and better connections at Wigan Wallgate. Improved services started on the Bury line at the same time, coincidental with the opening of the Whitefield Interchange. Rail services in the south of the conurbation were improved too, with better services on the Marple, Hadfield and Hyde loop lines.

By June 2,300 of GMT's 3,000 buses were equipped with two-way radios and a new control centre was opened at Atherton. The radios were of use not only in maintaining service regularity but in pinpointing trouble. Concern about increasing violence towards staff led GMT to start fitting sirens to buses, sounded by the driver pressing an alarm switch. The same concern had seen standard 7483 delivered in 1976 with a protective screen on the driver's cab. In the end the alarm sirens became a standard fitting; the protective screens did not.

A revenue protection unit – a sort of flying squad of ticket inspectors – was established in an effort to reduce fares evasion, estimated to be costing the PTE around £1 million a year out of revenue approaching £50 million. In its first year of operation it led to 85 prosecutions being initiated by GMT. Fares were increased in September by 11.5 per cent. The concessionary fare rose from 4p to 5p.

In November the new £3.5 million Tameside garage opened and replaced the depots inherited from Ashton and SHMD (at Stalybridge). This was the PTE's first new garage and it had a capacity of 180 buses. A small depot was retained at Glossop to which were transferred seven of the unusual short ex-SHMD Fleetlines, to replace ex-North Western buses. A noteworthy feature of the Tameside depot was the use of waste oil from its buses for the depot heating system.

Work was started on a new garage at Bolton, and on Bury Interchange. Hyde bus station was reconstructed. A new bus station in Manchester's massive Arndale shopping centre made progress during the year – but hit a snag when it was realised that there was only three inches of clearance between the roof of a standard double-decker and the ceiling of the bus station. This was considered inadequate for proper ventilation of the fumes emitted by the buses using it. It also posed problems in jacking a bus up in the event of an accident or a tyre failure.

There was more to GMT's bus stations than a steady flow of buses. A largely unsung operation was Metroflora – three flower shops located at the bus stations in Altrincham, Ashton and Manchester Piccadilly. Plans were in hand to extend Metroflora to Rochdale and Manchester Arndale – at the latter the sweet smell of fresh flowers would provide welcome relief from the overpowering aroma of diesel fumes. There were also seven bus station shops trading as Metrokiosk. The Metro branding on shops and florists had its origins in the 1974 plan to use Metro as a fleetname.

A significant event in December was the opening of the first railway station in Greater Manchester since 1938. This was at Brinnington, near Stockport, between Bredbury and Reddish North stations. Bus services 325/6 which ran between Stockport and Brinnington were revised to provide rail connections. There were recommendations to electrify the lines from Stockport to Hazel Grove (again – it had previously been suggested in 1972) and Manchester to Bolton. The Hazel Grove line was, in 1981. The Bolton line wasn't.

Selnec Parcels – the name had been retained by GMT – was still running at a loss, not having recovered from the ending of the GUS contract in 1973. Passenger loadings were still falling too, down to 455 million in 1976/77 from 483 million the previous year. The bus fleet totalled 3,026 – the last time it would top 3,000 – and the number of employees was 12,056.

75

1978: Integration of LUT

In 1978 the flow of new standards increased slightly. Chassis production at Leyland was now keeping up with demand and GMT was using two body suppliers – Northern Counties and Park Royal. The latter was bodying Atlanteans; the former Atlanteans and the start of a new order for 150 GMT Fleetlines, plus some for LUT. A total of 188 standards entered the GMT fleet – 143 Atlanteans and, starting with 8001 in June, 45 Fleetlines.

Orders were placed for a further 300 Northern Counties-bodied Atlanteans. The order for Leyland Titans (as the B15 had been christened) was increased to 120 and an order was placed for 50 more MCW Metrobuses. The 120 Titans grew out of the initial reservation made in 1976 for 50, the first of which were due in 1978. Also due were 30 Metrobuses from MCW (increased from the original ten) and the PTE's first new Dennises, two Dominators to be bodied by Northern Counties. The Dominator, launched in 1977 when Dennis reappeared in the bus market after a ten year absence, featured a Gardner 6LXB engine and a Voith automatic gearbox.

However all was not well with the manufacture of new models. The integral Titan, being built in London at Park Royal, was caught up in a skills dispute at the factory and production was running well behind schedule. Leyland did well to finish a GMT example in time for the 1978 Motor Show, the first at the National Exhibition Centre in Birmingham. This was 4001, which was exhibited on the Leyland stand. Northern Counties showed standard Atlantean 8212.

Unaware of the scale of Leyland's problems GMT allocated fleet numbers 4001-4035 with matching registrations ANE1-35T for the first Titans – and placed a follow-on order for 70 more, making a total of 190. MCW was faring little better with production of its new Metrobus, to the first of which GMT had allotted fleet numbers 5001-5030 with registrations ABA1-30T. Only four of the Titans arrived early enough to qualify for a T-suffix registration; none of the Metrobuses did. Unused marks ANE21-35T were allocated to service vehicles.

The two Gardner-engined Dominator chassis were delivered and stored at Charles Street, Stockport, towards the end of the year; one had earlier been displayed at the Motor Show by Dennis.

LUT received a further seven 11.3m Leyland Nationals (530-36) in drab all-over red. It also got a further five Plaxton-bodied Leyland Leopard coaches (537-41). New coaches were delivered to Charterplan too, in the shape of four Duple-bodied Leopards, one of which was in Warburton livery, and to Godfrey Abbott in the form of five Plaxton-bodied AEC Reliances. These started a new coach numbering series at 1. The Leopards were 1-4; the Reliances – the last AECs to be purchased by GMT – became 5-9 when Godfrey Abbott vehicles were allocated fleet numbers at the start of 1979.

During the year GMC decided that GMT should not pursue its interest in the Blundell group of coach companies and it consequently relinquished its right to acquire an 80 per cent shareholding.

Dial-a-Ride's ten new buses to replace its original fleet arrived in 1978-79. The eight Bedford CFs entered service in the spring and autumn of 1978 while the two Ford Transits arrived early in 1979. All had automatic gearboxes and 17-seat Reeve Burgess bodies. At 2184kg unladen the CFs were GMT's lightest buses.

LUT's new Nationals were to be its last red buses. By the spring of 1978 a total of 30 out of the order for 45 standard Fleetlines had been delivered to LUT in its attractive red and grey livery. When the last 15 (515-529) started to appear in August they wore a modified version of GMT's livery – orange with white relief between decks. The LUT fleetname was retained – but in GMT's standard lettering style and alongside the M-blem. The old-style Lancashire United name had vanished after 55 years unchanged use. The last routine red repaint was 1967 Guy Arab 282.

Above **A new contract for Northern Counties-bodied Atlanteans started at fleet number 8151 in 1978 and had reached 8229 by the end of the year. Two Wigan-based buses are seen in the town centre in the company of an ex-municipal PD2 Titan.** Stewart J Brown

Although Park Royal was unable to produce integral Titans, it was still delivering conventional bodies on Atlantean chassis to GMT. Two 1978 buses, out of a contract for 160 delivered between 1977 and 1979, load in Manchester Piccadilly Gardens. *Stewart J Brown*

A new coach numbering series started in 1978 with the delivery of ANA1-9T, four Charterplan Leopards and five Godfrey Abbott Reliances. Charterplan Leopard ANA1T arrives at Wembley Stadium on a private hire; it has a Duple Dominant II body. The Godfrey Abbott Reliances had Plaxton Supreme bodies with a break in the waist moulding to accommodate the upswept stripes. The deeper moulding used on Duple bodies was left intact and the stripes stopped at waist level on vehicles painted by the coachbuilder but were carried over the moulding on coaches repainted by GMT. *G R Mills, R Marshall*

In 1978-79 the Sale dial-a-ride fleet was modernised with five Bedford CFs and two Ford Transits. All had automatic gearboxes and Reeve Burgess Reebur 17 bodies. This is a Transit. *R L Wilson*

77

The livery selected for ex-LUT Bristols and Seddons was overall orange, as shown on 1974 Plaxton-bodied RESL 414 at Wigan. Stewart J Brown

The 1974 Show model Leopard, 424, received a non-standard livery layout with white window pillars and roof. It is seen in Swinton depot in 1981 by which time the LUT name had been dropped. R L Wilson

The layout of the single-deck LUT livery caused some problems. GMT single deckers were white with orange skirt and roof. None of LUT's single-deckers appeared in this style and early single-deck repaints appeared in either all-over orange, or orange with differing styles of white relief, namely: for the waistband (Seddons 390/92); for the top half of the bus (Seddon 386, Leopards 424, 450); for the window surrounds (Seddon 389) or for the roof.

In the end Nationals and Leopards were repainted orange with white roofs while other types were painted overall orange with no relief at all – an unusual decision for the PTE, which had always taken great care over the appearance of its fleet. In fact, the overall orange did not look too bad. Seddon 389 had its white window surrounds painted orange to make it all-over orange.

At first LUT crews refused to man the orange buses and the first did not enter revenue-earning service until the end of September at Atherton. At the same time Hindley agreed to accept the new buses, but Swinton did not fall into line for another week or so. In the interim Swinton was in the ludicrous position where crews refused to drive red Guy Arabs 168/9 because they had been drafted in to replace two orange buses.

Progress was made in switching LUT to Almex ticket equipment with omo conversions at Hindley and Swinton in June. LUT's fleet at this stage stood at 350 buses and coaches divided between Swinton (134), Atherton (126) and Hindley (90). On order for the LUT fleet for 1979 delivery were a further 44 standard Fleetlines (570-613) and 23 more Leyland Nationals (543-65).

Initially the single-deck livery layout for LUT buses gave some problems. Leopard 450 was one of three buses repainted in a traditional layout with the colours divided at the waist. It is seen in Wigan bus station in the company of Seddon 375, still in LUT red and grey. Both buses have Plaxton bodies. The standard LUT Leopard livery evolved as orange with a white roof, as illustrated by 455 in Atherton.
Stewart J Brown, M R Keeley

LUT Seddon 389 ran briefly in an attractive combination of orange with white window surrounds. It is seen fresh out of the paint shops at LUT's Howe Bridge workshops in Atherton.
J Robinson

Now that the bulk of the GMT fleet was in orange and white (apart from Wigan's renegade Titan 3230) the PTE felt able to indulge in the occasional commemorative livery. The first – and last – was 1963 Bury Atlantean 6316 which reverted to Bury's green livery and fleet number 116 to mark the 75th anniversary of the town's first electric tramcar. It was withdrawn in this condition and presented to the Greater Manchester Transport Museum later in the year – part of a plan to ensure the retention of a vehicle from each of the PTE's constituents.

Hattersley station on the Hadfield line was opened in May. Rail timetables were improved and local bus services revised to provide co-ordination with the trains.

May also saw the opening of Rochdale's new £4.3 million bus station which had 24 bus stands beneath a 570-space multi-storey car park. The bus station was on an island site and subways provided access to and from the central shopping area. The other move on the property front in 1978 was the closure in February of the tiny Ramsbottom garage at Stubbins Lane. Ramsbottom operations were transferred to Bury. The two depots had effectively been managed as a single unit from the early days of Selnec. Glossop was now GMT's smallest garage, 19 buses being allocated there.

Centreline service 5 began operation in Manchester in May as a Saturday shoppers service and was soon carrying 3,000 passengers each Saturday. It ran from 0900 to 1800 with a five minute frequency, connecting Piccadilly station, Piccadilly, Deansgate and Victoria station. A flat 6p fare (5p concession) was charged.

79

LUT's red and grey livery was abandoned in 1978 and as buses became due for repainting they received GMT orange. On double-deckers the livery layout which had been used on LUT's newest Fleetlines was adopted, and they were repainted orange with white between decks. The oldest repaints were some of the company's first Fleetlines, 1962 Northern Counties-bodied CRG6LXs. The longest-lived of these survived until 1982. M Fowler

The last of the front-engined Daimlers which had been acquired from Manchester City Transport reached the end of the road in 1978. All were Metro-Cammell-bodied buses new in 1963. One of the late survivors was 4637. R Marshall

A few of the surplus Seddon midis continued to find interesting new homes. Four were sold to the Greater Glasgow PTE in the spring, and were later joined by two (1735/6) on extended hire between 1979 and 1982. Another two (1730/3) were loaned to Ribble for operation in Clitheroe from August. They stayed with Ribble for two years, returning to GMT in October 1980. 1735/6 were repainted orange by GGPTE for operation on a city centre inter-station service; 1730/3 retained GMT livery during their sojourn with Ribble. Older-style Seddon 1704 fared less well and was converted into a mobile inspectors' bus and allocated to the revenue protection unit, as part of GMT's continuing clampdown on fare evasion. It was joined in 1979 by 1725. Greater Glasgow also borrowed a GMT National, 158. It ran in Glasgow from November 1978 to April 1979, returning south when GGPTE took delivery of its own first Nationals. Another bus to make a foray was standard Atlantean 7731, which was used in the spring on a European tour by Motorola with the upper deck equipped as a showroom.

The last front-engined Daimlers in GMT operation were withdrawn in November. There were six, 4632/36-39/41, former Manchester VM-registered CVG6s with Metro-Cammell bodies which had been new in 1963. They had survived at Princess Road and Northenden, reprieved after withdrawal two years earlier because deliveries of new standards had fallen behind schedule.

The oldest standards were coming up to their first major overhauls to coincide with the expiry of their initial seven-year Certificates of Fitness and all was not well with the Northern Counties bodies. The wide pillar spacing meant that the location of the body pillars did not coincide with that of the chassis outriggers. To compensate for this the body structure incorporated short vertical pillar sections at the outriggers. However these created unforeseen stresses in the structure which necessitated major rebuilding of the lower part of the body to incorporate a strengthening solebar. The rear bulkheads also had to be rebuilt. The body movement had also affected the rubbers which held the windows in place and small retaining clips were fitted to the pillars to stop the windows falling out under pressure – most notably from unruly schoolchildren. Park Royal's body structure differed from Northern Counties' and standards with Park Royal bodies did not suffer from these problems.

Transfers of vehicles acquired from the PTE's constituent operators slowed down as new standards made up an increasing proportion of the fleet – the 1,000th was delivered during the year – and the only transfers of note involved ex-North Western Bristol REs. One of the Marshall-bodied RESLs was moved from Stockport to Bury, while three went to Glossop in exchange for Alexander-bodied RELLs. The Bury RESL was for the Summerseat service on which passenger demand had outgrown the Seddon midis. A tight turning point in Holcombe Village precluded the use of Nationals.

Following the abandonment of the Picc-Vic link the PTE turned its attention to an alternative connection between the area's southern and northern rail networks. This became known as the Castlefield Curve, and Parliamentary powers were sought for its construction which attracted opposition from Manchester City Council. Unlike the Picc-Vic line the Castlefield Curve did not offer the prospect of attractive new city centre stations. Castlefield lay to the west of Manchester's central business district and the link was to run from Salford station via Castlefield junction to Deansgate station.

Fares were increased in July by 11.7 per cent and the concessionary fare was raised from 5p to 6p. GMT was investigating other sources of income and to try and raise extra advertising revenue ten standards were fitted with Sounds-in-Motion tapes. These played music and – for a maximum of nine minutes an hour – adverts. The tapes played only on the top deck; passengers who preferred to travel in silence could seek haven in the lower saloon. They initially operated at Tameside and Wigan and a passenger survey showed that only three per cent of users had any objection to the system.

A new style of so-called T-shaped offside exterior advertisement was becoming increasingly popular. These incorporated display material which fitted on the blank panel by the staircase and sat untidily on GMT's livery where they cut across the orange relief band at upper deck floor level. This prompted a rethink of the livery layout and two alternatives were tried in the autumn. Standard Atlantean 7599 appeared in October in orange up to the waist, white for the lower deck window surrounds and between-decks panels, and orange for the upper deck window surrounds and roof. It was accompanied by standard Fleetline 7328 in overall orange with the white relief applied only to the upper deck window surrounds and roof. Both liveries could accommodate T-shaped adverts but neither was deemed acceptable, although 7328 was in fact a pointer to the future. Rather than repaint the two buses, the non-standard liveries were used as a base for GMT adverts.

The parcels operation was still losing money (£70,325 in 1977/78) and its future finally came under detailed scrutiny. Discussions were entered into with the trade unions and it looked as though the operation would be sold. Instead the fleet was cut from 108 to 90 vehicles – and a small profit was made in 1978/79.

Passenger figures were down four per cent at 436 million and the bus fleet was now 2,987 vehicles operated by a staff of 12,181.

Top In the early 1970s LUT bought three batches of 33ft-long Daimler Fleetlines with Northern Counties bodies, totalling 26 in all. Most were 76-seat dual-door buses, but there were five single-door 79-seaters including 403 seen here on a Bolton local service. The L-shaped advert for Lancashire United Travel obscures the fleetname which was carried on the panel behind the driving compartment. Stewart J Brown

Centre Atlantean 7599 was repainted in a non-standard layout with a broad area of white between decks in an attempt to find a way of accommodating so-called T-shaped offside adverts without them cutting across breaks in the livery. Although not adopted for the fleet, rather than repaint the bus GMT used it to promote Saver Seven tickets. J Robinson

Fleetline 7328 was repainted orange with white for the top deck window surrounds and roof. Later it too received Saver Seven advertising.

1979: Late Buses and Service Cuts

The winter of 1978-79 saw widespread industrial unrest throughout Britain – the so-called winter of discontent. Strikes at GMT affected service reliability and caused further passenger losses. This was exacerbated by a strike by fuel tanker drivers which cut fuel deliveries and led to services being curtailed.

Bus deliveries were running late again and Leyland's decision to close its Park Royal factory called into question the future of the Titan. GMT's first, 4001, had been delivered at the end of 1978; the next three did not arrive until April and May, the fifth followed in September. By the end of the year there were only ten in stock – over 12 months after delivery of the first. They were divided equally between Oldham and Birchfields Road, with the first entering passenger service in August. One Titan, 4004, got to Manchester by way of Helsinki where it had been exhibited by Leyland at a UITP congress.

MCW was later in starting deliveries, but when Metrobuses began arriving they did so rather more quickly than Titans. Metrobus demonstrator TOJ592S was inspected (but not used) at Stockport in July. GMT's first ten Metrobuses arrived in October and November and were allocated to Oldham and Princess Road, with five buses at each depot. All had Gardner engines, Voith gearboxes and hydraulic brakes, which was in essence London Transport's specification. The option of GKN-SRM transmission was not pursued.

Above **The ranks of front-engined buses were being thinned down and the last ex-Ashton PD2 Titans were withdrawn in 1979. They had Roe bodies. One loads in Victoria bus station, while another lays over.**
J Robinson

Left **Ten Leyland Titans entered service with GMT in 1979 and they were intended to be the start of something big: 190 of the Park Royal-built integrals were on order. Oldham had five, two of which are seen here in the town centre. They were Gardner-engined 73-seaters.**
Stewart J Brown

GMT had received 10 out of 190 Titans on order and was in fact only going to get another five. It quickly ordered an extra 80 Metrobuses from MCW, bringing the total order for Metrobuses to 160. It also decided to aim for a 15-year vehicle life (instead of 12½). Deliveries of standards fell from 188 in 1978 to 135 in 1979. Northern Counties delivered 59 Fleetlines and 44 Atlanteans. Park Royal's contract for 160 bodies on Atlanteans was completed with 7960 in July.

A further 20 Leyland Nationals were ordered to help cover the shortfall on new deliveries and these entered service in April and May as 186-205, rather appropriately carrying registrations ABA11-30T re-allotted from Metrobuses which had failed to appear on time. They were shared between Leigh (five), Rochdale (nine), and Altrincham (two), and Oldham, Birchfields Road, Weaste and Bolton which each got one. At the same time 11 withdrawn ex-Manchester Fleetlines of 1965 in the 4701-30 series were recertified for further service at Princess Road, which needed low double-deckers to pass under Partington Bridge following road resurfacing. A further 12 from the 4731-60 batch were also recertified. One was soon withdrawn, but 22 of these 47xx Fleetlines were transferred into Princess Road in the first half of the year from Weaste, Birchfields Road and Northenden, with Princess Road releasing Mancunian-style Fleetlines in exchange.

Above **To cover late deliveries of Titans and Metrobuses a further 20 Leyland Nationals were added to the fleet in the spring of 1979 and were allocated registrations which had originally been booked for the so far non-existent Metrobuses.** J Robinson

Left **In 1966 Rochdale had taken delivery of six Willowbrook-bodied AEC Reliances which were the most modern Reliance buses to operate for the PTE. Most were sold in 1979.** A MacFarlane

83

GMT had aimed for a 12-year operating life for its buses but delivery delays and declining passenger numbers (which meant less revenue) forced a rethink, with the new target being 15 years. Some buses in the fleet had already reached this age, including 6694, an Atlantean which had been new to Bolton in 1964 and was by 1979 one of the oldest buses in the fleet. It was withdrawn during the year. J Robinson

Former municipal buses were becoming less and less common as time passed. Among those to go in 1979 was ex-Oldham Atlantean 5146 with Neepsend body. It had originally been withdrawn in 1977, but had then been recertified to give two more years of service. R L Wilson

A Tayside Regional Transport Alexander-bodied Ailsa, WTS275T, was inspected in October. This was the second Ailsa to have been examined by GMT and it led to an order being placed – for one chassis. This was delivered to Northern Counties in December.

Despite delivery delays GMT was still able to hire vehicles to the South Yorkshire PTE which was suffering from a severe shortage of serviceable buses. All were Mancunian-style Leyland Atlanteans. The first went to Sheffield in December 1978, followed by eight in February and five in March. They were returned to GMT in the summer. Standard Atlantean 7919 was inspected by London Transport but not used by them.

The work of the revenue protection team formed in 1978 led to 600 prosecutions for fares-related offences and to the decision to appoint a second team. Fares were increased in July by 15 per cent (with the concession fare going up from 6p to 7p) and at the same time bus mileage was temporarily cut by around eight per cent because of the fuel shortage. The cuts, which generally meant fewer buses in the evenings and on Sundays, became permanent.

There were a number of moves on the property front. In January the PTE became the owner of the Central Station site at the request of GMC to ensure that there would be no piecemeal redevelopment of it. It was owned by Central Station Properties, which was a PTE subsidiary. Central Station had been closed in May 1969 and after a period as a car park was ultimately to reopen in 1986 as the G-Mex Centre.

January also saw work start on the Bury Interchange. The new Bolton depot opened in March, replacing three smaller depots. It was designed to be big enough to house the entire allocation of Bolton and Bury – but whatever plans there may have been to absorb Bury's operations they did not reach fruition and in 1983 work was put in hand to build a new, and arguably unnecessary, £5m Bury depot.

The £1.75m undercover bus station at the Arndale shopping centre in Manchester was opened on 24th September, the problems about its limited height clearance having been overcome by lowering the floor. It had a capacity of 900 departures a day and was used initially by 14 services, most of which had previously terminated in nearby Cannon Street (16/7, 60/1, 102, 163/5/7, 254, 264) or Victoria (8, 12). From elsewhere in the city centre came the 50 (which had used Albert Square) and the 100 (formerly in Piccadilly). Three LUT services (26, 32, 39 from Victoria) followed soon after. At the same time the reconstructed bus station at Eccles was opened.

The opening of the Arndale bus station gave rise to MACABUS – the Manchester Central Area Bus Operating Strategy. This was designed to extend to Arndale the routes from the west which used the dismal Greengate terminus (under the broad railway bridge beneath the former Exchange station). Routes arriving in Manchester from the east were extended to Victoria bus station. The motives behind MACABUS were good, giving better city-centre penetration, but severe traffic congestion caused havoc to schedules and forced a rethink.

The PTE had ceased direct retailing through its own bus station kiosks in January, opting instead for a licensing arrangement which passed their operation to private individuals. This spelt the end for Metroflora as a PTE operation.

After protracted negotiations with the trade unions GMT launched its ClipperCard in November. This was a pre-purchased discount ticket which gave travellers ten trips for the price of nine, and was cancelled by being inserted into an Almex M cancellor. GMT had ordered 3,000 cancellors from Almex in 1976 but union opposition had delayed their installation and use. Because the cancellors had been in store unused for so long there were frequent breakdowns.

To combat service unreliability an attendance bonus was introduced for drivers and conductors in November. It resulted in an immediate improvement in services. As well as smartening up on crew attendance, GMT decided to smarten up appearances too and during the year a new style of brown uniform was issued to drivers and conductors. It replaced the blue uniform which had been used since the inception of Selnec. The use of brown for uniforms initially did not include inspectors; they had to wait until 1981.

A new training centre was opened at Bennett Street, adjacent to the Hyde Road depot. A new standard livery for training buses was introduced in February – white with yellow on the between-decks panels. A new livery layout, inspired by that used on Charterplan coaches, was also adopted for the Parcels Express fleet. The parcels operation was extended in the summer in conjunction with White Star Carriers of Wigan, an LUT acquisition.

LUT was still a separate operation, albeit with its buses now in a modified version of GMT's orange and white livery. In 1979 it received 23 Leyland Nationals (543-65) and more standard Fleetlines.

Following the trial in 1978 of ten buses fitted with Sounds-in-Motion advertising tapes it was decided in 1979 that a further 100 buses would be fitted. One tape-equipped standard, Atlantean 7926, was loaned to the South Yorkshire PTE in July for evaluation.

Transfers of note were six 1966 ex-Manchester Fleetlines from Northenden to Altrincham in exchange for somewhat younger Leyland Nationals, new in 1975. Fleetlines 6348-50, ordered by Bury but new to Selnec, returned to their home town after a seven year stint in Rochdale. Four former Oldham Atlanteans, E-registered buses with Neepsend bodies, were moved to Bury in the autumn. The arrival of these vehicles released six ex-Ramsbottom PD3s which moved to Queens Road (one), Wigan (four) and Rochdale (one).

The last ex-Ashton PD2, C-registered 5442, was withdrawn at the start of the year. The end of operation of rear-entrance ex-Manchester buses came a step closer in November with the withdrawal of 1964 Metro-Cammell-bodied Leyland Titan PD2/37 3703, the last PD2 at Queens Road. However it wasn't the end of traditional Titans at the depot. As 3703 left Queens Road, ex-Ramsbottom PD3 6406 arrived from Bury.

Another significant withdrawal was the Seddon-Lucas battery-electric midi, EX62, which was added to the Manchester Museum of Transport collection. It had seen little service. The Museum of Transport in Boyle Street, adjacent to Queens Road depot, opened its doors to the public in May and attracted almost 4,000 visitors in its first season. The Museum was a unique blend of official and enthusiast enterprise, being housed in a GMT-owned building but operated by volunteers from the Greater Manchester Transport Society, an organisation of local enthusiasts. It housed the PTE's own vehicle collection, representing vehicles acquired from Selnec's constituents, along with privately-owned buses and other transport items including uniforms, photographs and ticket machines.

Passenger figures fell by another four per cent, to 418 million. The fleet now stood at 2,930 and the number of employees at 12,197. This was the last time the employee figure would be over 12,000.

Among the buses taken out of service in 1979 was this ex-Bury East Lancs-bodied Fleetline. It is seen in Bolton on one of the services which linked the two towns. J Robinson

Leyland Nationals were added to the LUT fleet in 1979 and were in the new orange livery with white roofs. They were LUT's last Nationals and, like earlier deliveries, were 11.3m-long 49-seaters. One waits in Warrington to take up a trip to Wigan. J Robinson

1980: Rising Fares and Falling Patronage

The year started with a 14.6 per cent fares increase – less than six months after the previous one. The ClipperCard, launched in November 1979, was proving a success and ClipperCard prices were not raised when fares went up in January. Consequently weekly ClipperCard sales rocketed from 30,000 at the start of the year to 70,000 by March. Weekly sales fell back to 50,000 from March when ClipperCard prices were raised to reflect the earlier increase in cash fares.

ClipperCards offered bus users a discount and prices after the March increase were:

Single fare	10-journey ClipperCard
9p	£0.80
17p	£1.50
26p	£2.30
35p	£3.10

At the same time SaverSeven sales were averaging 50,000 a week and around 25 per cent of GMT's adult fare revenue was now coming from pre-purchased tickets.

Operations director Jim Batty, the former Bolton general manager, retired at the start of the year. His successor was Jack Thompson who had been Manchester City Transport's last general manager.

It took from November 1978 to May 1980 for Leyland to deliver 15 Titans out of GMT's orders for 190. The Titan was in trouble. Following the closure of Leyland's Park Royal factory because of poor productivity, Leyland announced that it would be switching Titan production to Eastern Coach Works. This was opposed by the ECW workforce on the grounds that it introduced unskilled labour to the factory and Leyland then decided to move the Titan to Workington.

Faced with this interruption to deliveries GMT had no option but to cancel its outstanding Titans and make do with 15. The Titans were allocated to Birchfields Road (4001-5/11), Oldham (4006-10) and Stockport (4012-15). Stockport's four had Leyland TL11 engines, making them the fleet's first turbocharged double-deckers. The TL11 was a development of the naturally-aspirated 680 engine used in GMT's Atlanteans. Of the remaining 11 buses all, apart from TL11-engined 4003, had Gardner 6LXB engines. 4003 was quickly moved from Birchfields Road to Stockport to bring all the Leyland-powered Titans together.

The Titan had Leyland's new Hydracyclic five-speed gearbox which had hydraulic rather than air operation and incorporated a retarder, a feature first seen in the GMT fleet on the two Foden-NCs. And, largely under the influence of London Transport, it had hydraulic rather than air brakes. Another novel feature was independent front suspension. This provided a better quality ride and improved passenger access. The Titans were delivered in an attractive LUT-style livery: orange with white relief between decks – but with the addition of a shallow brown skirt. They had standard GMT destination displays. Two of the registrations booked for Titans which never appeared, GNF16/7V, were used on standard Fleetlines 8141/2.

Recognising that Northern Counties supplied the bulk of GMT's double-deck bodies Leyland investigated the possibility of licensing Northern Counties to build Titans, but this plan came to naught. Indeed, officially Northern Counties was still planning to build the Foden-NC but the feedback from the six operators running the seven prototypes was not encouraging. No more were built.

Above Where Leyland failed with its new-generation Titan, MCW succeeded with its semi-integral Metrobus. GMT quickly built up a fleet of Metrobuses which started a new number series at 5001. 5045 shows the original LUT-style livery used for the Metrobuses. This view at Victoria Station in Manchester also emphasises the small wheelarches of the Metrobus, which was fitted with low-profile tyres. The use of a deep nearside windscreen was a carryover from the Metro-Scania and Metropolitan designs. *Stewart J Brown*

Opposite During 1980 the Northern Counties standard body was modified with the introduction of a different glazing method for the lower deck windows which were recessed slightly; the flush upper-deck window mountings continued unchanged. The polished body side mouldings at skirt level and upper deck floor level were deleted from the body specification in the same year. A 1980 LUT Fleetline, seen in Bolton after the 1981 fleet renumbering, shows the changes. *Stewart J Brown*

The two Dennis Dominators, ordered back in 1978, were delivered from Northern Counties in March and entered service from Birchfields Road in May. They were numbered 1437-38, carrying on the 14xx sequence from the two Foden-NCs. The standard body as fitted to the Dominator was immediately recognisable by the rectangular radiator grille on the front panel. Earlier types of rear-engined double-deckers had rear-mounted radiators (apart from the Metropolitans); the Dominator had the radiator at the front.

Delivery of MCW Metrobuses was also under way although after the first ten, 5001-10, in the autumn of 1979 there was something of a lull and it was not until June that more followed. A total of 24 out of the first 30 arrived before August and retained their booked GBU-V registrations (which had, of course, replaced an unused ABA-T series). Where the allocation of standard Atlanteans and Fleetlines had been (and continued to be) haphazard, the Metrobuses were despatched to depots in tidy groups of five. The first ten, 5001-10, were DR101/6 models with hydraulic brakes and were initially allocated to Princess Road and Oldham. The next twenty, 5011-30, were DR102/10s with air brakes and went in neat batches of five to Queens Road, Stockport, Tameside and Birchfields Road. All featured MCW's standard body with asymmetric windscreens and were in LUT-style livery. The Metrobuses were 73-seaters (43 up, 30 down), compared with 75 (43/32) for standard Atlanteans. They weighed 9566kg. The first of the third batch of Metrobuses, 5031, was displayed by MCW at the 1980 Motor Show and went to Queens Road in November.

The final six of the 1979 Metrobuses were allocated W-suffix registrations from a block, MNC486-550W, booked primarily for new standards:

MNC486-493W	Standard Fleetlines 8143-8150
MNC494-499W	Metrobuses 5018/9, 5021/3/6, 5030
MNC500-503W	Charterplan Leopard coaches 26-29
MNC504-550W	Standard Atlanteans 8304-8350.

Atlanteans 8304-8315 were originally to have been KDB304-315V.

New standards were still being delivered from Northern Counties and 110 were taken into stock in 1980. These included the final 46 from the 1978 order for 150 Fleetlines. The last, bringing to 500 the number of Selnec/GMT Fleetlines, was 8150 which entered service from Birchfields Road in August. There were, of course, 90 similar Fleetlines in the LUT fleet which were delivered over the period 1977-80. LUT's last Fleetline was 613, allocated to Swinton depot in December. A total of 64 standard Atlanteans was delivered from Northern Counties in 1980.

Detail design modifications during the year included a change to the body structure which involved the fitment of recessed lower deck windows. This affected LUT Fleetlines from 580; standard GMT Fleetlines from 8116; and Atlanteans from 8305. To cut costs the translucent roof panels were deleted from the body specification and the last bus to be fitted was 8334. Atlantean 8332 had one translucent panel instead of two, suggesting that someone in Northern Counties' purchasing department had failed to order an even number of these panels for these buses.

Rather late in the day GMT decided to try the front-engined Ailsa, ordering one. This appeared in 1980 with a Northern Counties body with a lengthened rear overhang. The front-mounted engine allowed a neat rear end but the high driving position limited the space available for the destination display. At first there was space for intermediate destinations, but this display was later removed. R Marshall

The polished side strips which marked the break between the orange and white paint at wheelarch and upper deck floor level were an attractive detail of the Selnec standards and had been perpetuated by GMT. But to cut costs these were also abandoned in 1980, from Fleetline 8133 and Atlantean 8298 (although 8299 had polished trim). LUT vehicles had in fact been appearing without polished trim since 1978, the last to be so fitted being 514.

On the plus side it had been decided in 1979 to fit moquette trim on both decks (the top deck had originally been trimmed in leather, but latterly a pvc-type material had been used). Fleetlines from 8101 and Atlanteans from 8229 had moquette trim throughout, as had LUT's 575 onwards (as well as a few earlier LUT 1979 Fleetlines). Subsequently seven-year-old standards coming up for their first Certificate of Fitness overhaul were also retrimmed with moquette seats upstairs. A new moquette pattern was adopted from 8488.

Variations on the standard theme appeared in the autumn, one of which was a pointer to the future and one of which was not. That which was not was 1446, the fleet's first front-engined Ailsa. 1446 was a Mark II model with a Volvo TD70E 6.7-litre turbocharged engine – the smallest power unit ever used in a GMT double-decker. It had a Self-Changing Gears gearbox. The chassis had in fact been built in 1978 (at the Volvo factory in Irvine) but had been delayed at Northern Counties as the coachbuilder adapted its standard GMT body structure to fit the Ailsa's chassis.

The Ailsa, with no engine intrusion at the rear of the lower saloon, was more capacious than rear-engined models and offered 79 seats (44/35) against 75 (43/32) on GMT's Fleetlines and Atlanteans. The high driving position and front mounted radiator of the Ailsa meant that the windscreens were higher than on a normal standard body, which in turn reduced the area available between decks for the destination display. 1446 featured a new style of display with a single glass panel behind which there was an offset route number alongside the final destination and a limited number of intermediate destinations. The display was still clear, but less informative than the Manchester City Transport-inspired layout which had continued through Selnec days and was still the GMT standard. The Ailsa was delivered in August and entered service at Stockport as a two-man bus in January 1981 after acting as a demonstrator at the 1980 Motor Show. An order was placed for four more (1447-50).

The pointer to the future was 1451, a prototype of Leyland's new Olympian. Even before the Titan project ran into trouble at Park Royal, Leyland had recognised there would be a need for a chassis which other bodybuilders – such as Northern Counties – could body. This was developed as project B45 and launched at the 1980 Motor Show as the Olympian. Such was the importance of GMT's business to Leyland that the fifth of six pre-production B45 chassis was built to a GMT specification.

The Olympian was built at Leyland Vehicles' Bristol factory and GMT's had a turbocharged Leyland TL11 engine and Leyland's Hydracyclic gearbox – a driveline inherited from the Titan. It also had air suspension and air brakes (the Titans had hydraulic brakes). The Olympian had a perimeter frame which was swept up over the front and rear axles and this prevented Northern Counties from fitting back-to-back seats over the rear wheels. Instead a three-passenger inward-facing bench seat was fitted, resulting in the loss of two seats in the lower saloon: the Olympian seated 73 (43 up, 30 down).

The Olympian was a low-frame chassis and this reduced the overall height of the Northern Counties standard body which cut the amount of space available for the destination display. Accordingly 1451 was fitted with an electronic dot matrix single-line destination display: high-tech but uninformative. It also had Firestone's large rubber HELP impact-absorbing bumper at the front. Neither of these features impressed GMT.

88

Of more importance was the new Leyland Olympian and a pre-production chassis was supplied by Leyland to Northern Counties and built to GMT specification. The body was again a lengthened standard – the Olympian was longer than the Atlantean and Fleetline which it was to replace – and it incorporated a front grille to provide a flow of air to the front-mounted radiator. It was displayed by Northern Counties at the 1980 Motor Show with an electronic destination display which was soon replaced by conventional roller blinds, as shown at Piccadilly Station.
Stewart J Brown, J Robinson

Both the Ailsa and the Olympian had a front-mounted radiator and the bodies fitted to 1446 and 1451 had grilles at the front, similar to those on Dennises 1437/38. Leyland demonstrated the Olympian to Grimsby-Cleethorpes Transport in December, but not in passenger service. As well as trying the Ailsa and the Olympian, GMT ordered two more Dennis Dominators to be fitted with air suspension. The pair of Dominators already in service had conventional leaf springs. An air-suspended Dominator in the shape of Leicester City Transport 206 (MUT206W) was inspected at Hyde Road in November. Two Scania double-deckers were also ordered.

89

Bury Interchange opened in 1980 and provided covered accommodation for waiting passengers. One of the East Lancs-bodied single-doorway Mancunians which were moved from Manchester to Bury in 1977 is seen leaving the Interchange followed by prototype standard 6397 (EX14) repainted in the new livery which was adopted at the end of 1980. *J Robinson*

The impressive Bury Interchange opened for business on 17th March, incorporating a new rail terminal to replace the former Bolton Street station used by trains from Manchester Victoria. This required the construction of a short section of new track. There was an escalator link from the rail platform to the bus concourse and the site also had a 105-space car park. A number of Bury area bus services were revised. The interchange cost £4.9 million and had taken 15 months to complete. A formal opening was held, with Princess Alexandra officiating, on 9th July.

Services in Mossley – the Mossley of SHMD fame – were recast in March following the first detailed household survey ever carried out into the transport needs of its citizens. This involved interviewing no fewer than 4,000 households and the new network was designed to increase revenue without increasing operating costs.

The first stage of a new £1.4 million bus station in Daw Bank, Stockport was opened in April, with the second phase following at the end of November. This replaced long-standing terminal points in nearby Mersey Square. The first phase of another major construction project, a new garage at Altrincham, was completed in September.

The Saturday Centreline 5 shoppers service was re-routed from the end of March to follow the same route as used by Centreline 4 on weekdays. As a result the number of Saturday passengers doubled to 7,000.

The operation of double-deck AECs by GMT ceased in March with the withdrawal of 1966 Renown 6905. This was also the last ex-Leigh double-decker in the fleet. AEC Reliance bus operation also came to an end in 1980 with the sale of the last of the ex-Rochdale 1966 buses, 6026. Reliance coaches survived in the Godfrey Abbott fleet. Another significant withdrawal in April saw the last ex-North Western double-decker depart. It was one of the distinctive Daimler Fleetlines with lowheight Alexander bodywork, 4207, and was new in 1966. Rear-entrance buses ceased operating from Tameside in October when two 1967 Stockport PD2s, 5845/51, were withdrawn. The last Salford PD2s also came out of service during the year.

By the spring of 1980 only two ex-Manchester PD2s remained: 3696 – still with Manchester fleet number transfers – was withdrawn during the year, but 3706 survived into 1981. Both were new in 1964 and had Metro-Cammell bodies. They are seen in Hyde Road garage in March 1980. *J Robinson*

90

LUT's coach fleet was based on Leyland Leopards with Pneumocyclic gearboxes but before ordering for 1980 it evaluated three demonstrators in February and March. One was the first Leopard to be built with a ZF S6-80 manual gearbox, VCW85V, a PSU3 fitted with a 53-seat Plaxton Supreme IV body. Alongside this LUT tried similarly-bodied Volvo B58 KNS401V, and Duple-bodied Volvo GHS7T. Despite LUT's old loyalties, the Volvos won and three Plaxton-bodied B58s joined the LUT fleet in 1980 as 614-6.

LUT still had 45 Guy Arab Vs in service at the start of 1980. The last open-platform Arab, 110, was withdrawn in June, but this still left a fleet of forward-entrance Guys which were not suitable for omo. Former Manchester Fleetline 4731, new in 1966 and thus marginally older than most of LUT's surviving Arabs, was examined to ascertain whether it would be worth putting elderly rear-engined buses into the LUT fleet to speed the omo conversion programme.

The answer was yes – but the buses came from London, not from Manchester. LUT bought ten London Transport DMS-type Daimler Fleetlines, the first of which entered service in June on the 26, linking Leigh and Manchester. These were the company's first second-hand buses since 1929. An order for a further ten immediately followed and all 20 were in operation by October. All were around seven years old. They were therefore newer than LUT's Arabs and could be expected to give LUT a further seven years of operational life. Surprisingly they were CRL6 models with Leyland 680 engines. The LUT fleet had no Leyland-engined double-deckers, and all of GMT's Fleetlines were Gardner-powered.

Most of the LUT DMSs had Park Royal bodies but four were Metro-Cammell-bodied. Sales of LT Fleetlines were being handled by Ensign Bus, the Purfleet dealer, and Ensign rebuilt the LUT vehicles from LT two-door layout to single-doorway. GMT-style destination displays were fitted by LUT at Atherton. The buses entered LUT service as 71-seaters (44 up, 27 down) and were numbered 318-337 below the oldest complete batch of buses still in LUT service, Seddons 338-357. The intake of 20 DMSs and 34 new standard Fleetlines by LUT reduced the fleet of Guy Arabs to nine by the end of the year. None of LUT's Guys received orange livery, although some much older Fleetlines – 97-99 of 1962 – did.

The first foreign chassis for LUT in modern times arrived in 1980 in the shape of a trio of Volvo B58s with 55-seat Plaxton Supreme IV bodies. These were the only Volvo coaches in the GMT fleet until 1985 when it bought its first B10M. G R Mills

There were still 45 Guy Arabs in LUT service at the start of 1980. The newest were 1967 models with forward-entrance Northern Counties bodies, one of which picks up in Swinton. None of the Guys received PTE livery, although older Fleetlines did. Stewart J Brown

Right **To speed replacement of LUT's Guy Arabs and allow more rapid conversion of services to one-man-operation, 20 London Transport Fleetlines were bought from Ensign Bus. Surprisingly all were Leyland-engined; both GMT and LUT ran Gardner-engined Fleetlines. Bodywork was by both Park Royal, as on this bus, and Metro-Cammell, and all were rebuilt with full LUT destination boxes and the removal of the centre exit door. 2320, seen leaving Wigan bus station after the 1981 renumbering of the LUT fleet, was originally London's DMS588.**
Stewart J Brown

Considerable pruning was going on at LUT, primarily with the aim of reducing its spare vehicle capacity to a level in line with that prevailing in the GMT fleet. LUT's mixed fleet had around 100 buses taken out of it in 1980 (offset by only 54 vehicles going in) and casualties included the 20 Marshall-bodied REs of 1967, the 20 Alexander-bodied Bristol REs of 1969, the three surviving Tiger Cubs and the first of the 1970 Seddon Pennine RUs, as well as all but eight of the Guys. Two of the Tiger Cubs, 242/4, had been operating on hire to Staffordshire independent Stonier of Goldenhill from November 1979 and were sold to Stonier in 1980 without returning to LUT.

The final stage of the integration of LUT was made known in the autumn: it was to be absorbed by GMT on 1st April 1981. LUT's last five Fleetlines, 609-613, were delivered in December with GMT fleetnames. LUT's three depots had earlier in the year been allocated GMT-style codes:

AN	Atherton
HY	Hindley
SN	Swinton

At the end of the 1979/80 financial year only 219 of the PTE's 2,882 buses were not suitable for omo – less than ten per cent of the fleet. But the actual rate of omo conversions was still lagging behind the availability of suitable buses. In March 1980, 69 per cent of GMT's mileage was omo, compared with only 60 per cent 12 months earlier. The target was 95 per cent.

By May the economic recession gripping the country (unemployment in Greater Manchester had almost doubled in 12 months and passengers were set to fall by 10 per cent) had placed the PTE's finances in a precarious position and an embargo was imposed on capital expenditure. Transport minister Norman Fowler visited the PTE in July to discuss its problems and a further fares increase, of 15 per cent, was implemented in August – the third in just over 12 months.

Two new ClipperCards were launched in August – Shop-n-Save for shoppers and TeenTravel for teenagers. Both were designed to increase off-peak travel and the TeenTravel card was also aimed at retaining young travellers as bus users when they ceased to be eligible for children's concessionary fares. A mobile ticket sales unit (ex-LUT Bristol RE/Plaxton 248) was used to promote sales in the west of Greater Manchester County.

On the rail front, work started on electrification of the Stockport to Hazel Grove branch line which included reconstruction of Hazel Grove station. Improved service patterns were introduced on the Manchester-Oldham-Rochdale lines. BR also published proposals for a new line between Windsor Bridge (Salford) and Deansgate – this became known as the Windsor Link and was to allow through working from lines on the south of the city to those heading north-west. The proposal was linked to BR's plans to electrify the route from Manchester to Preston, and was designed to avoid the cost of electrifying the lines into Victoria station.

In October there was a major bus service revision in the Wilmslow area which involved Crosville and saw further integration of services between Manchester and Macclesfield and Stockport and Wilmslow. One result was that GMT buses now ran through to Macclesfield.

October also saw the announcement that the Sale dial-a-ride operation which had been started jointly with the Godfrey Abbott Group in 1974 was to cease at the end of the year. Passengers on the dial-a-ride service had fallen by 20 per cent while maintenance costs had risen by 70 per cent. The operation had lost £24,000 in 12 months and needed new buses. In the end it received a last-minute reprieve and continued until the end of January 1981.

Top At the end of 1980 a new GMT livery appeared – orange with brown skirt and white upperworks. It was also to be applied to former LUT vehicles. LUT Fleetline 490 is in the rare combination of new GMT livery with old LUT fleet number; LUT buses were renumbered in April 1981. M R Keeley

Above On Metrobuses the new livery incorporated a deeper brown skirt which was taken to the tops of the wheelarches. 5056 leaves Oldham depot. Stewart J Brown

buses – 13 Titans and two Atlanteans – being redistributed amongst Manchester garages, leaving only six Titans in Salford. In the summer former Manchester Fleetline 4737 was moved to Oldham, along with Mancunian Atlanteans 1142/43 which stayed there for a year. Four further Fleetlines and three Mancunian Atlanteans operated on loan in Oldham in the autumn.

The four remaining ex-Leigh buses were on the move in the summer of 1980. Leopards 6060/61 ran in Stockport for a few months while the depot's Bristol VRTs were being recertified. They were then moved to Wigan in July to join sister buses 6062/63. All four were withdrawn in 1981.

Leopard/Plaxton coaches 69-75 were demoted to dual-purpose vehicles and repainted orange with a white roof.

The strong identity presented by the original Selnec-style livery was being diluted. The Metropolitans, Metrobuses and Titans were in non-standard liveries, as was the LUT fleet. And the T-shaped exterior adverts which had led to two experimental liveries in 1978 were becoming more widespread and did not sit comfortably on the Selnec livery layout.

All this forced a rethink on livery with the aim of producing a layout which could be applied to all body styles. The result appeared in November on standards 7264 and 7351 which featured orange as the main body colour but with a white roof and upper deck window surrounds and a brown skirt. The entrance doors were also white, but on dual-door buses the exit doors were painted orange and brown. The same livery was applied to 7328 which had previously been in an orange and white experimental livery, similar to the newly-adopted standard but without the brown skirt.

The brown skirt was the same depth as the orange skirt had been on the original Selnec livery. It was intended to apply the brown to a much shallower area on the new generation Titans and Metrobuses but only a few Titans received the shallow skirt when they were repainted; none of the Metrobuses did. The single-deck version was orange with a brown skirt and white roof and was first applied to Leyland National 113. This was an interesting bus in its own right, being an amalgam of two accident damaged Leyland Nationals, 113 and 168, which were rebuilt as a single bus at Charles Street works during the year. 113 had extensive front end damage; 168 was damaged at the rear. The highly-standardised manufacturing methods used at the Leyland National plant in Workington ensured complete interchangeability of parts and made the job of joining the good parts of two damaged buses relatively easy. The use of brown on the vehicles echoed the PTE's corporate use of the colour in new bus stations and on uniforms.

Top **On the single-deck version of the new livery the white relief was restricted to the roof. A 9ft 11in high Leyland National squeezes under the 10ft bridge at Middleton Junction. 1337 was originally EX37, one of the eight 11.3m Nationals purchased in 1972.** Stewart J Brown

Above **Ex-Wigan PD3 3230 avoided the GMT orange paintbrush and was repainted in 1980 in its former municipal livery, complete with Wigan Corporation fleetname and Wigan-style fleet number transfers. New in 1961, it was not withdrawn until 1981. It had a Massey body.** M R Keeley

The imminent withdrawal of the dial-a-ride service paled into insignificance against an announcement made at the end of November. As GMT continued to grapple with its financial problems, service cuts were to be made across the entire operation with a target saving of £3 million. Around 2,400 jobs would be lost. The reality was slightly less harsh – but 2,000 jobs did go over a four-year period.

As the GMT fleet became increasingly standardised the transfer of interesting pre-PTE buses became less frequent. The year started with no fewer than 15 ex-Salford

GMT's management structure with three divisions in Manchester/Salford and eight districts for the rest of the Greater Manchester area was reviewed in 1980. The outcome was the formation in spring 1981 of four areas, each headed by a manager. Passenger figures were down again, to 411 million while the number of employees dropped by almost 500 to 11,720. But bigger cuts were to come.

1981: Restructuring and Cutbacks

In January the cheerful-sounding Southern Cemetery bus station and the £650,000 Oldham Town Square bus stands were opened. Stockport bus station was officially opened in March. Work also started on Wythenshawe bus station which was completed at the end of the year and cost £630,000. The new garage in Altrincham was completed in June. Who could have foreseen the changes which would lead to its closure ten years later?

Market Street in central Manchester was pedestrianised in February; bus services were re-routed via High Street, Cannon Street and Cateaton Street. The small bus terminal in Cannon Street near the delightfully-named Hanging Ditch, disappeared at this time.

Above **Among 1981's new standards was 8353, an Atlantean allocated to Rochdale which was part of the newly-created North area. Although a new livery had been devised at the end of 1980, new buses continued to be delivered in the old colours.** J Robinson

Four area general managers took office in April as part of GMT's management restructuring. The new area structure was:
- North, with 700 buses at Bury, Rochdale, Frederick Road, Queens Road and Weaste was managed by Dennis Rodgers who had been running Charterplan and had previously been at LUT. Rodgers' place at the coaching operation was taken over by Jim Hulme.
- South, with 680 buses at Northenden, Princess Road, Stockport and Altrincham was under the control of Tom Dunstan who had been district operations manager and had come to the PTE from North Western back in 1972.
- East was the biggest area with 785 buses at Hyde Road, Birchfields Road, Tameside and Oldham and was managed by Ian Bradshaw, previously the Central South division general manager.
- West, with 750 buses, covered Leigh, Wigan, Bolton and the three LUT depots which were absorbed into the west area when LUT ceased operating as a separate company at the end of March. Ken Holt, who had been LUT's general manager, became West area manager.

A voluntary redundancy scheme was announced in January and by March, as the new area structure was being put into place, cuts were being made in staffing levels as the recession raised unemployment and reduced the demand for transport. This was all part of a major series of service reductions started in 1980 which by 1982 would see the fleet cut by 10 per cent, bus mileage reduced by nine per cent and staff cut by 16 per cent. During 1981 some 400 buses were taken out of service, against an intake of 230 new vehicles.

Over six years passenger numbers had declined by almost 30 per cent and the PTE reviewed its accommodation requirements in the west, examining the garages at Bolton, Salford (Frederick Road and Weaste), Queens Road (Manchester), Swinton, Atherton, Hindley, Leigh and Wigan. It proposed that three depots – Hindley, Leigh and Weaste – be closed, but the plan was not implemented in the face of strong opposition from the unions and local politicians.

The Hale Barns executive service, an early Selnec innovation which dated back to November 1970, was withdrawn in March. It had lost £11,000 in 1980 despite an extensive publicity campaign.

The dial-a-ride minibus fleet at Sale was absorbed into the main GMT fleet in February after the service ceased. The minibuses were moved to Altrincham where they were available for private hire. Six were renumbered 1743-1748 (following on from the Seddon midis). The remaining four were withdrawn.

As an experiment fares on services using the Ashton New Road in Manchester were reduced in April to assess the effects of fare cuts on short-distance travellers and on off-peak long-distance travellers. Launched as the "Great Spring Fare Saving", initially for three months, the experiment lasted 12 months and showed that the fares reductions, which averaged 38 per cent, generated around 10 per cent more passenger journeys – but that this was insufficient to cover the cost of the cuts.

On 1st April Lancashire United Transport with 299 buses and coaches was absorbed by GMT and incorporated in the West area. The 299 vehicles included eight Guy Arab Vs, ten Bristol REs, 38 Seddon Pennine RUs, 20 ex-London Transport DMSs and three Volvo B58 coaches along with Leyland Leopards, Leyland Nationals and Daimler Fleetlines.

The LUT fleet had been numbered in a single series irrespective of vehicle type. This, of course, was not the GMT way so the fleet was partly renumbered. LUT's coaches were allocated numbers 30-51, continuing the Charterplan series which had reached 29. LUT's Leyland Nationals joined the GMT series with numbers 206-250. There was a mixed assortment of other single-deck buses in the LUT fleet – Seddon Pennine RUs, Bristol REs and Leyland Leopards. All of these retained their existing numbers except for three whose numbers clashed with ex-North Western REs. The range of numbers occupied spanned 338 to 464, with LUT Seddon RUs 338, 339 and 341 being renumbered 347 to 349, clear of the former North Western buses (which finished at 344). Withdrawn LUT Seddon 342 was reinstated in passenger service in May with new fleet number 354. LUT's double deckers were mainly Daimler Fleetlines and the oldest were given numbers from 2305 (after the last Mancunian). Most fitted into the 23xx and 24xx series by simply adding 2000 to their LUT numbers. The eight remaining Guy Arabs were allotted numbers 2451-2458. LUT's 90 Fleetlines to GMT standard specification were numbered in a single block, 6901 to 6990, fitting below the original Selnec standard series commencing at 7001.

The new livery solved the problem of how to paint ex-LUT single-deckers – all gradually received the standard GMT scheme as they became due for repaint. A 1976 Plaxton-bodied Leopard loads in Atherton. Stewart J Brown

Some of the newest ex-Wigan buses, long-wheelbase 1972 Atlantean AN68/2Rs 3330-3339, received the brown-skirted livery, including 3333 seen in the town centre. The body is by Northern Counties. These buses were taken out of service in 1984-85. Stewart J Brown

The new GMT livery which had appeared at the end of 1980 was applied to buses, including those acquired from LUT, as they were due for repaint, replacing both the original 1970 Selnec layout and the 1978 LUT scheme. But it did not feature on new Atlanteans being delivered from Northern Counties until August, starting with 8414 – although 8415/16 emerged from Northern Counties' paintshops in the old livery. 8416 was the first standard Atlantean to be allocated to a former LUT garage (ironically in old-style GMT livery) and was sent to Hindley. By the end of the year four new Atlanteans were running from ex-LUT garages.

The new livery was not applied to any of the few remaining half-cab buses in the fleet.

Few ex-municipal buses received it either; one which did, after an accident, was Oldham Atlantean 5180. Other interesting recipients of the new colour scheme in 1981 were three 22xx Mancunians from the SVR-K batch, ex-North Western Marshall-bodied Bristol REs 291, 299 and 301 which were based at Bury (for services to Holcombe Brook and Summerseat) and one Metropolitan, 1426.

Above **Mancunians were on the way out when the new livery was introduced but a few were repainted including 1971 Park Royal-bodied Fleetline 2226 which was operated until 1984. It is seen on the so-called circular 53 route which ran around the eastern and southern suburbs of Manchester from Cheetham Hill in the north to Old Trafford in the west but was not actually a complete circle.** P Sykes

Left **Only a few ex-North Western buses received the 1980 livery. These included Marshall-bodied RE 301, seen operating out of Bury depot.** Stewart J Brown

The PTE had for most of its life been aiming for a vehicle life of around 12 years and the 12th year of its existence saw the departure of the last buses from some of its original municipal constituents. In January the last Ramsbottom buses were withdrawn, Titans 6409-11, which had been running in Wigan for just over a year. 6411, the last home-market PD3, was repainted in Ramsbottom livery at Wigan depot and passed to the Greater Manchester Museum of Transport. The last SHMD buses in the fleet were withdrawn in October. They were short 1968 Northern Counties-bodied Fleetlines 5640/2/7. The same month saw the last Rochdale buses go, 1964 DK-registered Fleetlines 6223/5. The last Leigh bus, 1968 East Lancs-bodied Leopard 6061, was withdrawn in November.

Another withdrawal of note was 1354, one of the two Mercedes-Benz buses added to the fleet in 1973. It came out of service at Oldham in November.

Top **Wigan depot had acquired the remaining ex-Ramsbottom Titans which by this time had been fitted with route number displays. The three newest, 6409-11, were withdrawn in 1981 and 6410, new in 1969, passes 1967 bus 6407 which was withdrawn in 1980. They have East Lancs bodies.** M R Keeley

Centre **1981 saw the demise of GMT's few ex-municipal Leyland Leopard service buses, with the withdrawal of ex-Leigh 6060-64 and ex-Stockport 5084-88. All were PSU4s with East Lancs bodywork. 6064, one of the ex-Leigh vehicles which had been re-allocated to Bolton, is seen in Bury. The Leopards were the last ex-Leigh buses in the GMT fleet.** M R Keeley

Below left **This ex-Daimler demonstration Fleetline was taken out of service in 1981. It had dual-door Willowbrook body and was delivered to Selnec in Rochdale colours – Rochdale had committed to buying it shortly before the PTE was created. It ended its days in Bolton.** Stewart J Brown

Below right **The PTE's only AEC Swifts – 14 in number – were at Rochdale and were withdrawn in 1980-81. 6034 was one of four Pennine-bodied examples taken over from Rochdale Corporation and dated from 1969.** M R Keeley

97

Standard Atlanteans still dominated the new vehicle intake with 141 being delivered in 1981. This was a marked increase on 1980's 110, and was the highest number yet achieved by Northern Counties. The two Scania chassis for GMT were delivered to Northern Counties in May, as the PTE's engineers continued to evaluate alternative bus models.

The Atlanteans were accompanied by a further 89 Metrobuses, 5032 to 5120, all of which were delivered in the mainly orange Metrobus livery. Most of the Metrobuses had white doors (which was the GMT standard) but 30 (5016-20, 5031-35 and 5041-60) had orange doors. Deliveries were spread throughout the year and the buses were still allocated to depots in neat batches of five. Metrobuses were introduced to Hyde Road in 1981 and added to the fleets at the six original Metrobus depots – Oldham, Stockport, Tameside, Princess Road, Queens Road and Birchfields Road. The Metrobus allocation at the end of the year was:

Altrincham	10
Birchfields Road	15
Hyde Road	20
Oldham	10
Princess Road	10
Queens Road	20
Stockport	20
Tameside	15

The Altrincham Metrobuses were the first ten, 5001-10, which were transferred from Princess Road and Oldham. These differed from the rest of the Metrobus fleet in that they had hydraulic brakes. All Metrobuses from 5011 had air brakes.

The Metrobuses were not without their problems. New gearbox mountings were developed by GMT's engineers to stop gearboxes from falling out. The modified mounting design was later adopted as standard by MCW. Potentially more serious was the problem of engines falling out. With what seems like a touch of high farce, GMT's engineers had to install a microswitch in the engine bay which detected any untoward movement and triggered a warning light in the driving compartment to let the driver know the engine was loose. MCW eventually came up with a cure.

Northern Counties changed their construction methods during 1982, introducing a new alloy-framed body which was similar to the previous steel-framed structure but had thick rubber-mounted gasket glazing and a new one-piece glass fibre staircase. The change took place in October. Atlanteans 8433 and 8437/8 were the first GMT buses built to this design which was then used for all buses from 8443. The alloy-framed design addressed the structural weaknesses of the original steel-framed body.

In July 1981 GMT had 2,540 buses divided between 20 operational locations as follows:

Depot	Code	Allocation
Altrincham	AM	69
Atherton	AN	94
Bolton	BN	178
Birchfields Road	BS	160
Bury	BY	99
Frederick Road	FK	128
Hindley	HY	68
Hyde Road	HE	187
Leigh	LH	46
Northenden	NN	134
Oldham	OM	169
Princess Road	PS	204
Queens Road	QS	198
Rochdale	RE	125
Stockport	ST	207
Swinton	SN	99
Tameside	TE	142
Glossop	TE	19
Weaste	WE	90
Wigan	WN	124

The Charterplan coach fleet (incorporating Godfrey Abbott) was housed at Charles Street, Stockport (the former North Western Road Car headquarters) and coaches in Warburton livery were kept at Bury.

Olympian 1451, which had been retained by Leyland for proving work after the 1980 Motor Show, entered service from Queens Road in October, still with its Firestone rubber front bumpers and electronic destination display. It was the last bus to enter service in the original orange and white Selnec livery layout.

The two air-suspended Dennis Dominators arrived in October and were initially sent to Birchfields Road where they joined the first two, before being moved to Oldham in November. They were numbered 1439/40, consecutive with the first Dominators and had 75-seat (43/32) Northern Counties bodies, Gardner 6LXB engines and Voith D851 automatic gearboxes.

A Leyland National 2 demonstrator, GCK430W owned by Fishwick of Leyland, was tried for three weeks in July and August. It was allocated to Stockport and used on crew-operated duties because it lacked ticket-cancelling equipment. However GMT had but a limited role for single-deck buses and no National 2s were ordered. Old single-deck types to vanish in 1981 included the AEC Swift. Only 14 had been operated, all either ex-Rochdale or ordered by Rochdale. Two had been taken out of service in 1980; the remaining 12 followed in 1981. This brought AEC bus operation to an end, although Reliances remained in the coach fleet until 1984. The single-deck Fleetline accompanied the Swift to oblivion with the withdrawal of the survivors of the 13 which had been operated. Twelve had come from Bury and Rochdale; the thirteenth was an ex-demonstrator which Rochdale had made a commitment to buy.

GMT was the only PTE still running half-cab buses in 1981. However as efforts were made to cut costs omo had leapt forward and now accounted for 84 per cent of GMT's operations, compared with only 69 per cent in 1980. The last ex-Manchester Titan was withdrawn in April. It was 3706, a 1964 VM-registered Metro-Cammell-bodied PD2 which had outlasted the rest of its class by 12 months and had even acquired ticket cancellors for ClipperCards. To achieve this its seating capacity was reduced by two on the upper deck and one on the lower deck. The last Wigan Titans, 3230/3 (which were 20-year-old PD3As) and 3280 (a 1966 exposed-radiator PD2), came out of service in November. PD3A 3230 went to the museum.

The PTE's first new Dennises arrived in 1980 in the shape of two Dominators. They were followed by two more in 1981, one of which was 1439, seen here in Oldham which housed all four from 1982. Like the Leyland Olympian, these buses had front-mounted radiators which required the provision of a small grille at the front. Stewart J Brown

Above **Nine F-registered ex-Wigan Titans lasted into 1981. Massey-bodied PD2 3291 loads in the town centre.** M R Keeley

Right **The remaining front-engined Titans were nearing the end of the road and by the end of 1981 the last few were running from Stockport. They were the survivors from 27 PD3s delivered to Stockport Corporation in 1968-69. The last of these, 5897, was one of five with forward-entrance East Lancs bodywork. It is seen in Manchester Piccadilly in March 1981 and was withdrawn later that year. A few of its contemporaries survived into 1982.** M R Keeley

As GMT's half-cabs slowly disappeared (a few survived into 1982) the first wholesale withdrawal of Mancunians commenced. All but one of the 1968 Atlanteans and Fleetlines were withdrawn in 1981 and a start was made on the 1969 models too. In the LUT fleet, 21 Seddon RUs from 1970-71 were taken out of service. Wigan's Panthers were slowly being withdrawn and this led to the appearance of Nationals in GMT's Wigan fleet. Five 1974 13xx-series Nationals were transferred from Altrincham in March. Other buses on the move included four ex-Salford F-registered Atlanteans from Frederick Road to Bolton and half-a-dozen North Western RELLs which made their way back to Stockport after a spell at Princess Road.

Between April and June nine standard Fleetlines in the 73xx series were hired to the South Yorkshire PTE to cover a vehicle shortage. Further Fleetlines, including a pair of newer 80xx-series buses, were sent to South Yorkshire between November 1981 and the spring of 1982.

The 22 ECW bodies supplied on Leyland Leopards in the early 1970s had been less than successful. They had structural problems at the rear because the design had been developed for the rear-engined Bristol RE and had not been adequately re-engineered for the mid-engined Leopard. And they were functional rather than stylish. Consequently the decision was taken to rebody eight of them, 82-89, which would rid the fleet of rather unattractive coaches and allow Charterplan to re-equip at around half the cost of buying new vehicles.

The old bodies were removed at Charles Street works and the chassis were then dispatched to Duple at Blackpool to be rebodied for the 1982 coaching season. The ECW bodies were sold to a dealer, Martins of Middlewich, and one actually saw further service with enterprising Lancashire independent Holmeswood Coaches, who fitted it to a 1965 ex-Midland Red Leyland Leopard chassis.

A new coach livery was introduced at the end of the year with simple horizontal bands of colour at skirt level replacing the angled stripes used since 1976. Duple-bodied Leopard 2 was an early Charterplan repaint; Plaxton-bodied AEC Reliances 5 and 9 were the first Godfrey Abbott repaints. Godfrey Abbott Bedford YRT 24 was transferred to the Warburton fleet, running with Warburton names on its Godfrey Abbott green livery.

Uncertainty still surrounded the parcels operation and it was transferred on 1st April to a new company, Parcels Express, owned jointly by the PTE and United Carriers. In its last year as a wholly-owned PTE subsidiary it had lost £158,971.

June saw the start of electric operation on the Stockport to Hazel Grove railway line. A different sort of electric operation was being examined for one of the Seddon midibuses. Battery power had been abandoned and the two experimental battery buses were out of service. Instead Seddon 1733 was chosen for an experiment with dual power sources – until it was found to be suffering from some mechanical defects when 1729 was used in its place.

The existing engine and gearbox of 1729 were removed and a small diesel engine was mounted at the rear. This was a three-cylinder 2.1-litre Hatz air-cooled Silentpack, and it was used to provide belt drive to electric generators rated at 32kW. The Hatz engine ran at a constant 3,000rpm and was able to power the bus up to 30mph. Regenerative braking was included and the whole package upped the weight of the bus from 4569kg to 5250kg, an increase equal to the weight of about seven passengers.

The aim was to reduce energy consumption by 20-25 per cent and when the bus was unveiled in the summer it was scheduled to enter service in November or December; this was later put back to May 1982. In the event 1729 never entered service.

Top **A shortage of buses at South Yorkshire PTE led to the hiring-in of GMT Atlanteans. 8166 is being driven into Sheffield's Pond Street bus station in a spirited manner in August 1981 with an uninformative destination display offering a choice which includes Marple and Romiley – which just happen to be on GMT's route 383... presumably the bus-load of passengers knew where they were going. 8166 carries the GMT M-blem, but has had its fleetname removed.** G R Mills

Centre **After trying battery power in the 1970s the PTE looked at the possibility of a hybrid diesel-electric bus in 1981. Seddon 1729 was fitted with a three-cylinder rear-mounted Hatz diesel engine which drove electric generators to propel the bus. The livery was a non-standard variant of that used on the Seddon midis, with the addition of a brown skirt. 1729 was not used in service after its diesel-electric conversion.** J Robinson

Above **The PTE's coach livery was revised at the end of 1981. The upswept stripes which had been introduced in 1976 were replaced by two bands of colour along the body side. The basic livery was grey for the lower panels and white for the upper part of the body. The broad strips of colour at wheelarch level were the same as those previously used for the upswept stripes – two-tone blue in the case of this Leopard in the Warburtons fleet. The end result was attractive and restrained.** R Marshall

There were no fares increases in 1981 for the first year ever and the PTE announced a fares freeze – although this was to be short-lived. Cheap Sunday Rover tickets were introduced in August at 80p for an adult or £1.75 for a family ticket. These were sold on buses and were valid for travel throughout Greater Manchester and into parts of Derbyshire and Cheshire. Their use covered not only GMT buses (except on airport express route 200 and the 349) but also services within the Greater Manchester area operated by Ribble, Crosville, Trent (at that time all NBC subsidiaries), Mayne and the West Yorkshire and Merseyside PTEs.

To encourage bus use by students a £37.50 term ticket was launched, giving unlimited travel for one academic term on the Oxford Road/Wilmslow Road corridor between Whitworth Street and West Didsbury. This proved a success – but its introduction was partly motivated by GMT's desire to keep competition out. Coach operator Finglands had applied for a licence to serve the university area. This was successfully opposed by GMT, but clearly forced a closer look at marketing tactics.

At the end of the year the range of ClipperCards was updated. The Shop-n-Save variant was rebranded as the Off-Peak ClipperCard and a new Concessionary ClipperCard was introduced.

A major rebuild of Mancunian 1066 saw it appear in the guise of GMT's Exhibus – a mobile exhibition unit which incorporated a cinema in the upper saloon. The conversion was carried out at the Charles Street works in Stockport.

Wholesale withdrawal of Mancunians started in 1981 when most of those delivered in 1968 and 1969 were taken out of service at the end of 12 years operation. 1066 saw further use with GMT as a mobile exhibition unit and was rebuilt for its new role at Charles Street works during 1981. Stewart J Brown

Another conversion in 1981 centred on standard Park Royal-bodied Atlantean 7032 which lost its roof in an argument with a low bridge and was converted to open-top as a replacement for ageing ex-Stockport PD2 5995 which dated back to 1951 and was withdrawn in 1981. The conversion was carried out at Frederick Road but was not completed until 1982. GMT hired a Merseyside PTE open-top PD2, CWM153C which had originally operated for Southport Corporation, to cover the need for an open-topper to perform duty as the Piccadilly Radio Fun Bus in December.

Passenger figures took a real dive in 1980/81, dropping 10 per cent from the previous year's 411 million to 371 million. The official fleet strength was 2,813, down slightly, but there was a big loss of jobs over the year – over 1,000 had gone, representing a nine per cent cut, leaving a work force of 10,682. More cuts were to come.

The changing face of GMT. Standard Northern Counties-bodied Atlantean 8411 in Middleton was one of the last buses to be delivered in the original Selnec-style livery. It is followed by a Metrobus in the predominantly orange livery worn by early deliveries, and a Park Royal-bodied Atlantean repainted in the new brown-skirted colour scheme. Stewart J Brown

101

The end of the Titan operation in Greater Manchester came in May 1982 when the three surviving ex-Stockport Titans were withdrawn. One of these was 5891 with rear-entrance East Lancs body.
Stuart Burton

1982: The Last Half-cabs

Withdrawals featured prominently in 1982 with some milestones being reached as the newest of the buses acquired in 1969 from Selnec's municipal constituents approached the end of their working lives. The year also saw the end of half-cab double-deckers.

The last ex-Ashton buses, 1969 PTF-G-registered Atlanteans 5459/60, were withdrawn in April. These were followed in May by the final ex-Bury bus, 1969 Atlantean 6392. In August the surviving buses from the fleets of Salford (1969 PRJ-G-registered Atlanteans 3156/7), Oldham (1969 SBU-G Atlanteans 5179/82) and Bolton (OBN-H Atlanteans of 1969) came out of service. The final Ashton-style Atlanteans, 5462-65, which were ordered by Ashton but delivered to Selnec, were also withdrawn in 1982.

However these withdrawals did not bring to an end the presence of GMT's original municipal founders. Apart from the Manchester-designed Mancunians there were still the last vestiges of orders placed by some of the smaller municipal constituents of the PTE. These included Atlanteans ordered by Oldham and Fleetlines ordered by Bury which survived into 1983. The last of the TWH-K-registered Atlanteans ordered by Bolton remained in service until 1985.

The last of LUT's half-cab Guy Arabs (2456-58 of 1966) were withdrawn at the end of January. The remaining ex-Stockport buses were withdrawn in May. They were also GMT's final half-cabs and the last 'deckers in the fleet with manual gearboxes – PD3 Titans 5891/4/6. Of these, 5891 had a rear-entrance body; the other two were of forward-entrance layout. With them came an end to over half a century of traditional Titan operation in Greater Manchester.

The last ex-Ashton buses were withdrawn in 1982. These were five Northern Counties-bodied Atlanteans of 1969 vintage. A batch of similar buses, ordered by Ashton but delivered new to Selnec, also came out of service during the year.
M R Keeley

More surprising withdrawals included the entire batch of MCW Metropolitan double-deckers after only seven years in service. The Metropolitans were heavy on fuel and the bodies were corroding badly. GMT was not the only operator to withdraw Metropolitans rather than invest in expensive repair work for their seven-year recertification. A start was also made on taking the Metro-Scania single-deckers out of service. Four had gone by the end of the year and the remaining nine followed in 1983 – but they had at least managed ten years service. The two unsuccessful Foden-NCs, 1435/36, were also withdrawn in 1982 after only six years operation: 1435 came out of service in February following a mechanical failure; 1436 followed in August. The surviving Mercedes-Benz, 1355, was also withdrawn in August.

The brown, orange and white livery was spreading through the fleet and among the more unusual buses to receive it were a further six 22xx-series Mancunian Fleetlines from the PNA-J, RNA-J and SVR-K registration batches, making a total of nine, and ex-Wigan long-wheelbase AN68/2R Atlanteans in the 3330-39 series. The Mancunians had been in overall advertising liveries on which the contracts had expired.

Metrobus deliveries recommenced in April with 5121, the first MCW to be delivered in the new livery, and finished in September with 5170. As before they were sent out to depots in tidy batches of five; all went to existing Metrobus depots. Standard Atlanteans with Northern Counties bodies continued to arrive and 135 were delivered during the year; those from 8526 were of the uprated AN68D specification.

The two original Dennis Dominators, 1437/38, were transferred from Birchfields Road to Oldham in April to join 1439/40. All GMT's Dennises were thus once again at one depot.

A new image was adopted for the 200 Airport Express service, operated by Northenden using 1973 Plaxton-bodied Leopards 70-75. To coincide with improvements to the timetable, coach 70 was repainted in April in GMT brown, orange and white with the addition of a blue band lettered 'Airport Express'; identical Leopard 71 was similarly repainted in June. This was a watered-down version of a plan to repaint the coaches mid-blue with a dark blue skirt and white roof, but did have the merit of retaining a strong link with GMT's corporate orange and brown. In May Northenden received four ex-Lancashire United 1977 Leopards, 35-38, to replace the older GMT coaches; 37 found its way on to the 200 in LUT livery.

The limitations of the electronic destination display fitted to Olympian 1451 had quickly become apparent and in the spring of 1982 it was fitted with conventional destination blinds – but showing a reduced display similar to that first seen on Ailsa 1446. This was to become the new standard and was used on subsequent deliveries of Olympians from 1983.

Orders were placed for a further 20 Metrobuses, to be 5171-90 and for 25 Leyland Olympians with Northern Counties bodies, to be 3001-25. The engine order for the Olympians was divided between Leyland, who were to supply ten TL11s, and Gardner, who were to provide fifteen 6LXBs. The 6LXB was already in use in large numbers in GMT's Fleetlines and was the standard power unit for the Metrobus. The turbocharged TL11 was something of an unknown quantity, only being fitted in the GMT fleet to the Olympian prototype and five of the Titans. GMT's experience of running Leyland 680s and Gardner 6LXBs

A major change in the structure of the Northern Counties standard body had come in 1981 when an aluminium alloy frame was used in place of the previous steel frame. The alloy body was immediately recognisable by its rubber-mounted recessed windows on both decks. Atlantean 8541 was a 1982 delivery to Wigan. M R Keeley

All change. In 1978 Victoria Station's facade was drab and grimy and the GMT service to the airport was simply route 200 being operated by an ECW-bodied Leopard in predominantly orange livery. By 1982 the station facade had been restored to its former glory and the 200 had been branded as Airport Express with a broad blue waistband on the side of Plaxton-bodied Leopard 70. The coach was in fact two years older than the ECW-bodied vehicle it had replaced, but the overall effect was of a marked improvement in service.
M R Keeley/Stewart J Brown

103

Left **Two more Ailsas (out of an order for four) entered service with GMT in 1982. They were allocated to Wigan and had a reduced destination display with no intermediate details. The Ailsa's high driving position precluded the fitment of the standard GMT three-aperture display.** Stewart J Brown

Right **The rather plain ECW-bodied Leopards, as illustrated above by 87, had been transformed. Eight were sent to Duple for rebodying and emerged with Dominant IV bodies and new X-suffix registrations. This vehicle was allocated to the Godfrey Abbott fleet.** M R Keeley

showed that the Gardner used less fuel and had a longer life between overhauls. There was also a reluctance to introduce the complication of turbochargers if they could be avoided and naturally-aspirated Gardner engines thus became the GMT standard in Olympians, no doubt to the disappointment of Leyland who had supplied many more engines to the PTE than had Gardner.

The first of GMT's production Olympians, 3001 with TL11 engine, was at the 1982 Motor Show. It was billed as the 1,250th Manchester standard body from Northern Counties and the livery incorporated white lining at skirt level. It entered service from Northenden at the start of 1983. The body on this and subsequent Olympians differed from previous standards in that it featured a thick pillar for the first bay on the upper deck. The Olympian had a longer front overhang than the Atlantean and this arrangement allowed the continued use of standard sized windows.

GMT had in 1980 ordered four more front-engined Ailsas, to be 1447-50. Two of these were delivered in June 1982 as 1447/48. These had 79-seat Northern Counties bodies and were allocated to Wigan; 1447 was crew-operated while 1448 was an omo bus and each had a different destination layout. The remaining two were cancelled. Their chassis were bodied by Marshall and they ended up in Glasgow with the Strathclyde PTE, an enthusiastic user of the Ailsa.

Seven of the eight 1975 Leopards which were being rebodied arrived from Duple in December 1981 and January 1982 with new Dominant IV coach bodies. 82/3/7 were allocated to Charterplan; 84/5 to Warburton; and 86/8 to Godfrey Abbott. All had new matching SND-X registrations and were in the new coach livery. The eighth was being equipped with a wheelchair lift for the carriage of disabled travellers and did not appear until 1984. Bristol coach operation ceased in 1982 with the sale of the last of the three 1975-76 LHSs.

Twelve Mancunian-style Atlanteans in the 11xx-series were transferred to Wigan from Manchester and Salford, along with four similar 12xx-series buses based on a Salford City Transport order. Most were withdrawn in 1983.

The long-running saga of the decline and fall of the parcels operation came to an end in January when for £75,000 United Carriers purchased GMT's 76 per cent shareholding in the joint company formed nine months earlier. United Carriers also purchased Parrs Wood depot.

The 1981 fares freeze came to an abrupt end in 1982. Greater Manchester was one of a number of Labour-controlled local authorities which sought to improve the attractiveness of public transport by keeping fares down. Throughout 1981 fares were pegged at the level they had been fixed at in August 1980. However the principle of freezing fares was successfully challenged in a battle, which was well publicised, between the Labour-controlled Greater London Council and the Conservative London Borough of Bromley and as a result GMT felt that it had no option but to increase adult fares to avoid falling foul of the law – which was still unclear. This it did with reluctance. Adult fares went up by 15 per cent on 4th April and at the same time the concession rate went up from 9p to 10p. The 15 per cent rise was broadly in line with inflation over the 18 month period of the fares freeze.

A new one-day leisure ticket, the Peak Wayfarer, was introduced by the PTE in conjunction with British Rail and other bus operators and was valid on buses operated by GMT from the Manchester area into the Peak District and North Cheshire. It cost £2.25 adult, £1.50 child/senior citizen, or £4.50 for a family ticket. The Peak Wayfarer could be purchased in advance and validated by the holder on the day chosen for travel. It was valid from 0900 on weekdays and all day on Saturdays and Sundays.

On 31st May Pope John Paul II visited Manchester and all traffic – except buses – was excluded from an area two miles around the Pope's Heaton Park venue. Special shuttle bus services provided by GMT linked Heaton Park with car parks and rail stations in surrounding towns. Normal services in the area were suspended. The Papal transport operation involved around 1,000 buses from all GMT depots. Souvenir tickets were issued to travellers on the shuttle services.

The operation was not the roaring success which had been expected. On the evening of 30th May a fleet of buses was stationed in Aytoun Street in central Manchester to take to Heaton Park people who wanted to sleep out to be sure of a place the next day. Few did. And on the day there were many more buses than were needed as the expected crowds failed to materialise – dissuaded perhaps by extensive advance publicity of the crowds which were expected. The operation also posed potential problems with drivers' hours regulations which GMT was prepared to try to get round by describing the papal visit as an emergency – a weak argument for an event which had been planned for 12 months.

Service revisions in the Rossendale area in June saw GMT buses reaching Burnley for the first time when the 473 from Bury to Rawtenstall and Water was re-routed and extended across the moors.

Work on a £1.1 million reconstruction of Ashton bus station commenced in February. Plans were also announced for the reconstruction of Radcliffe bus station, the redesign of the bus station at Moor Lane, Bolton, and for new bus stations in Leigh, Wigan and at the expanding Manchester Airport.

The Basic Bus Network Survey was started in March in the Urmston, Flixton and Davyhulme area. This was designed to examine the minimum acceptable level of bus service and to develop a long term bus/rail network strategy.

The Pope's visit to Manchester in 1982 saw a large area around Heaton Park closed to traffic with GMT called in to provide transport for visitors. Mancunian 2300 heads a line-up of almost 40 buses at Heaton Park. R Marshall

GMT was one of a number of major operators to support the Bus & Coach Council in its opposition to changes in the road service licensing system being floated by the government. Four new Atlanteans carried the enigmatic message 'We'd all miss the bus' and 8585 heads south past Princess Road depot, its home base. J Robinson

The venerable open-top PD2 acquired from Stockport was pensioned off in 1981 and its replacement appeared in 1982. Park Royal-bodied standard Atlantean 7032 was converted to open top after hitting a low bridge. Stewart J Brown

GMT's 100 double-deckers with taped music in the upper saloon were temporarily silenced when Sounds-in-Motion, the company responsible for selling the advertising and producing the tapes, went out of business. But for the few passengers who objected to top-deck music the respite was brief and the tapes restarted in November when a new company, Audio Bus Advertising, took over. The buses involved were all at Queens Road and Princess Road depots.

The first stirrings of government on the road towards bus deregulation were made in 1982 and the bus industry's trade association, the Bus & Coach Council (BCC), started lobbying on behalf of its members – including GMT – who were in general opposed to the abolition of the road service licensing which protected their local monopolies in the provision of bus services. To support the BCC's campaign four new standard Atlanteans – 8581/2/4/5 – entered service in November in the BCC's "We'd all miss the bus" livery and sporting the BCC logo which featured three stylised passengers sitting behind a driver – although to the uninitiated it bore more than a passing resemblance to a line of performing seals balancing beach balls. They were initially allocated to Frederick Road, Tameside, Swinton and Stockport respectively but the intention was to move them between depots and in January 1983 they were re-allocated to Weaste, Oldham, Hindley and Northenden.

An excess charge of five times the normal fare became payable by passengers caught trying to avoid paying their fares after 4th January. This was designed to reduce the PTE's lost revenue – now estimated at as much as £2 million a year out of a turnover of around £80 million – and passengers had the choice of paying on the spot or within 21 days. The introduction of the scheme was agreed with the trades unions but one, the Transport & General Workers, subsequently changed its mind when it appeared that drivers might have to implement the charge. In the event excess fares were collected by inspectors. The effectiveness of the revenue protection unit was underlined with over 1,000 prosecutions for fare evasion during the 1981/82 financial year. A Festive Fare offer was launched on 6th December and continued until 3rd January 1983. It fixed the maximum off-peak fare on buses at 46p, and a maximum off-peak rail return fare of £1.

Management changes in 1982 saw Ken Holt being replaced as West area manager by Ian Bradshaw, who was in turn replaced as East area manager by Jim Hulme. Hulme had been running Charterplan, where he was succeeded by David Verity.

At the end of GMT's financial year – 31st March – the fleet stood at 2,590, down from 2,813 a year before, a reduction of eight per cent. The spares percentage was cut from 22.4 to 21.1. Staffing levels had fallen sharply for the second successive year, down eight per cent to 9,806 from 10,682. In three years the PTE had shed 2,400 jobs. Passenger journeys fell to 353 million.

Radio control was extended to the entire fleet, managed from a new £140,000 communications centre at Devonshire Street North. The control rooms at Bury, Tameside and Atherton were closed.

1983: Olympian Arrival

The April 1982 fares increase had made short-distance travel relatively expensive and in February 1983 fares for journeys for distances between one and 2.5 miles were reduced by up to 34 per cent. For a 1.5 mile trip the fare came down from 35p to 23p. A new Travel Companions Club was launched in April, designed to encourage travel by the over-50s.

Off-bus ticket sales continued to do well. The SaverSeven was selling at the rate of 30,000 a week; the Saver Monthly at the rate of 3,000 a month. A Saver Annual was introduced. Not only did the pre-sold Saver tickets cut boarding times, they offered cash flow benefits to GMT with passengers paying in advance for their travel. And in a further attempt to stimulate travel GMT teamed up with BBC Radio Manchester in a joint promotion to encourage trips by bus.

As well as receiving 86 new standard Atlanteans (which had reached number 8693 by the end of the year), GMT continued to try experimental types. In March, two Scania BR112DH double-deckers were delivered with 75-seat Northern Counties standard bodies. These, 1461/2, were allocated to Leigh which was still running the surviving Metro-Scania single-deckers from 1972-73. The last of these were withdrawn in October. The BR112 had an 11-litre Scania DN1104 engine (a development of the engine used in the Metro-Scania and Metropolitan) which was rated at 193bhp and coupled to Scania's own automatic gearbox. It had of course air suspension, power steering and a retarder, *de riguer* on any up-to-date 1980s urban bus. In common with earlier Scania designs it retained the radiator in the unusual position ahead of the rear axle. The Scanias had spigot-mounted wheels which were painted grey rather than brown to show that they were not interchangeable with the wheels on other types of double-decker.

The final outstanding Metrobuses, 5171-90, were delivered in May and June and five each were allocated to Queens Road, Birchfields Road, Princess Road and Tameside.

Delivery of Leyland Olympians from Northern Counties commenced in April, running alongside delivery of Atlanteans. Like the Metrobuses, these were allocated to depots in blocks of five. Buses up to 3018 were in stock by November; 3019-25 followed in the first quarter of 1984.

Leyland-powered Olympians 3002-5 went to Northenden (to join 3001), followed by 3006-10 at Stockport and 3011-15 at Bolton. Of the first Gardner-powered examples 3016-20 were allocated to Frederick Road while 3021-25 went to Bolton for comparison with the Leyland-engined buses. 3017-25 had turbocharged 6LXCT engines; 3016 (and all subsequent Olympians) had the naturally-aspirated 6LXB. Like the Scanias, the Olympians had spigot-mounted wheels and these were painted grey. Olympians 3001-25 were to have been registered ANA1-25Y, but buses from 3011 entered service after August and had A-prefix registrations. The DVLC's decision to retain numbers 1-20 for future sale meant that 3011-20 did not have registrations to match their fleet numbers. The final Olympian, 3025, had a higher driving position. This was adopted as standard on subsequent deliveries.

Above **Delivery of the first 25 Olympians started in 1983 and five with Leyland TL11 engines were allocated to Stockport where they ran alongside similarly-powered Titans. Leyland saw the TL11-engined Olympian as the natural successor to the Atlantean, although in the end GMT opted for Gardner power, perhaps with an eye to the latter builder's factory being in GMT's operating area – and Gardner's reputation for economy and durability. To maintain the use of standard size window glasses the Northern Counties body was revised on production Olympians to incorporate a thick pillar after the first upper deck side window. On GMT's prototype Olympian 1451 the extra length had been added to the rearmost upper deck window. The Olympian chassis frame was lower than the Atlantean and Fleetline and this reduced the overall body height – once again forcing GMT away from its standard destination layout because of a lack of space above the windscreen.** M R Keeley

Northern Counties at Wigan – in the PTE's operating area since the boundary change in 1974 – had become the sole supplier of standard bodies to GMT since the closure of Park Royal in 1980. It had in any event been the major supplier since the formation of the PTE and its close involvement in developing the standard body. In the early 1980s it shared GMT's custom with MCW who were supplying complete Metrobuses. Not only was Northern Counties the principal supplier to GMT; GMT was conversely Northern Counties' biggest customer. In June it was announced that the PTE was purchasing the company, a move which effectively gave it an in-house bodybuilder, although Northern Counties continued to operate independently and to supply bodies to an increasing range of operators.

Northern Counties provided two new bodies for old chassis in 1983. Former LUT Fleetline 6912, new in 1978, had had its body burned out in a fire at Hindley garage in August 1981 and emerged with a new 75-seat alloy-framed standard body in January. It was followed in April by older long-wheelbase ex-LUT Fleetline 2413, burnt out at the same time. This had originally been fitted with two-door Northern Counties bodywork to LUT design; it re-entered service with a lengthened 81-seat (47/34) version of the GMT standard steel-framed single-door body. 6912 was thus the only Fleetline in the GMT fleet to have alloy-framed rather than steel-framed standard bodywork.

Two non-standard double-deckers were delivered in 1983. These were Scania BR112s allocated to Leigh which had experience of Scanias through its association with the Metro-Scania single-deckers which were being withdrawn as the two new buses arrived. The BR112s had 75-seat Northern Counties bodies. The two are seen in Leigh bus station. *Stewart J Brown*

Rebodying of damaged buses was not a feature of Selnec or GMT. However two relatively modern Fleetlines were damaged by a fire at LUT's Hindley depot in 1981 and were sent to Northern Counties to have new bodies fitted. 2413 was a long-wheelbase CRG6LXB/33 and it was returned to GMT service in 1983 with a lengthened version of the steel-framed standard body with 81 seats – six more than on a conventional standard. It is seen at Atherton depot, alongside a Leopard still in LUT red and grey. *R Marshall*

107

Other LUT buses in the news were the 20 ex-London Transport Fleetlines, five of which were repainted in standard GMT brown/orange/white livery and all but one of which had their engine shrouds removed.

In March the ten Gardner-engined Titans, 4001/2/4-11, now a non-standard type, were moved to Glossop from Birchfields Road and Oldham. The five Leyland-engined buses remained at Stockport. The solitary Ailsa at Stockport, 1446, was moved to Wigan in April to join the other two Ailsas in the fleet.

The 55 standard Park Royal-bodied Fleetlines delivered in 1973 (7151-7205) had been allocated to the Selnec Northern and Southern companies; none went to Central and it was not until 1983 that one found its way to a Manchester depot, when 7199 moved from Oldham to Birchfields Road.

The last ex-North Western buses in the fleet, Bristol REs 291/9 and 301, were withdrawn in September. They had been operating at Bury where they were replaced by newer ex-LUT Bristols 418-420, transferred from Swinton. These ran in Bury until 1985 and 419 received standard GMT brown-skirted livery. Withdrawal of 1972 Leyland Nationals in the 133x series started in 1983 and Selnec's first, 1330, was added to the collection in the Manchester Transport Museum. It was originally numbered EX30 in the experimental series and was the first production National. The last Bury-style Fleetlines, ordered by Bury but delivered to Selnec in 1970, were also withdrawn in 1983. More significant was the withdrawal of 5468-70 and 6396. These were the first of the prototype standards to go. They had been new in 1971-72 as EX3-5 and EX13.

The Charterplan coach operation had a strong upmarket image, reinforced in 1983 by the delivery of two new Setra S215HD integrals. German manufacturer Setra was widely recognised as the builder of the highest-quality coaches available in the UK at the time. They were the fleet's first rear-engined coaches and entered service as 30 and 31, carrying registrations 515VTB and 583TD transferred from ex-LUT Guy Arabs in the training fleet which were re-registered BNC988/9B. The Setras were powered by 14.62-litre 280bhp Mercedes-Benz OM422 engines, making them easily the most powerful vehicles in the fleet.

Top Withdrawal started of the prototype standards in 1982 although some survived until 1984, including Rochdale Fleetline 6252 with dual-door Northern Counties body. It is seen awaiting custom in Rochdale's modern bus station, sited below a multi-storey car park. *Stewart J Brown*

Centre and right Four Setras delivered in 1983 were Charterplan's first rear-engined coaches. The Setra was widely recognised as the premier European touring coach and the four S215HDs acquired by Charterplan marked a distinct move upmarket. The first two carried registrations transferred from ex-LUT Guy Arabs which were in the training fleet. Setra 583TD on tour at Wells, Somerset, contrasts with Guy trainer 583TD photographed at Atherton in 1981. The Guy was new in 1962 and had a 73-seat Northern Counties body. *M R Keeley, R L Wilson*

With the Setras came Charterplan's first Leyland Tigers. There were four with Duple Goldliner high-floor bodies, further evidence of Charterplan's upmarket image. Numbered 21-24, all were 46-seaters and, like the Setras, had toilet compartments. 24 was in Warburton livery and operated as a 38-seater with tables for Bury FC. All of 21-24 were delivered by March. They were followed by further Tigers in the summer, 52/3, which had Duple's attractive new Laser bodywork. These were allocated to Godfrey Abbott and Warburton respectively. All of the Tigers had 218bhp Leyland TL11H turbocharged engines and air suspension. Orders for 1984 comprised two more Setras, four Plaxton-bodied Leyland Tigers and a Leyland Royal Tiger Doyen rear-engined integral. The two additional Setras, 32/3, actually arrived in October; the Doyen did not.

To cater for the growth in traffic at Manchester Airport standard Atlantean 7638 had its lower-deck seating capacity reduced to 15 and operated alongside the Leopard coaches normally used on the service. In its new form 7638 seated 58 with space for luggage, compared with 49 seats in the Leopards. It wore a double-deck version of the Airport Express livery with a blue band between decks. A new £520,000 bus station was opened at Manchester Airport in December.

Improvements at Middleton bus station were completed in June, although it was only used by northbound buses because of a one-way street system which would have added to the mileage of southbound vehicles. Work started on a £263,000 refurbishment of Radcliffe bus station in August. The reconstruction of Ashton bus station was completed.

Although the bus industry was still regulated by the Traffic Commissioners, it was open to operators to seek licences to run local bus services. Old-established independent operator Mayne had retained its bus operations in Manchester throughout the period since the PTE had been formed. However a new operator appeared in August, in the shape of Citibus Tours, running a shoppers service from Middleton to Harpurhey five times daily, Monday to Friday, and eight times on Saturday. Citibus was to expand later in the decade.

Towards the end of the year GMT announced a bold new project to operate ten dual-mode diesel electric buses which were to be built by Shelvoke & Drewry, best known as makers of dustcarts. The work was being co-ordinated by a company called Hybrid Vehicles Ltd and followed the none-too-successful trials with Seddon midibus 1729 which hadn't actually turned a wheel in revenue-earning service after its dual-mode conversion. The programme was to be partially-funded by the EEC to cover the cost of £600,000 for GMT's ten S&D buses plus ten vans (to be operated elsewhere). The electric vehicle development group of University College Swansea was to evaluate the project. But, like so many electric vehicle projects this one never got off the ground. Diesel buses continued to rule the roads.

The first transfers involving ex-LUT buses took place in the autumn with 11 Leyland Nationals moving from Swinton to Wigan while six went from Atherton to Leigh. In exchange standard Fleetlines from Leigh and standard Atlanteans from Bolton were moved to Atherton. Five of Bolton's TWH-K registered long-wheelbase PDR2 Atlanteans were transferred to Manchester's Hyde Road depot in October.

Free travel was provided on GMT bus services on Christmas Day and Boxing Day. In the run up to Christmas a Festive Fares promotion allowed off-peak ClipperCards to be used on evening peak hour services from 5th December. This offer continued until 2nd January 1984. But the gilt was rubbed off the Festive Fares promotion by a series of four one-day strikes at most depots in December during productivity negotiations with the unions.

The downward trend in key statistics continued. Passenger figures had dropped to an all-time pre-deregulation low of 342 million. A decade earlier the figure had been 523 million without the inclusion of Wigan and LUT. In real terms passenger figures had about halved since the PTE's formation. The fleet strength was 2,530 with a staff of 9,730, both figures slightly down on the previous year.

Above **The first examples of Leyland's new Tiger for the PTE's coaching operations were delivered in the winter of 1982-83 and had Duple's short-lived high-floor Goldliner body. 24 was allocated to Warburtons and was used as the Bury Football Club team coach.** R Marshall

Below **Two more Leyland Tigers were added to the coach fleet later in 1983, and had Duple's attractive new Laser bodywork. 52 carried Godfrey Abbott livery.** G R Mills

1984: Light Rail Plans

Street-level light rapid transit was in vogue in 1984 with a 60-mile network being proposed by the joint GMC/BR/GMT rail strategy study which had been set up in 1982. The network would run on five existing and one re-opened rail route with on-street running in Manchester city centre between Deansgate, Victoria and Piccadilly stations. Two-car articulated vehicles would be used. The scheme would cost £90 million. The study group considered guided buses – but rejected a guided busway system as being too expensive.

The five lines earmarked for light rail were Altrincham, Bury, Oldham/Rochdale, Hadfield/Glossop and Marple/Rose Hill. The Didsbury line, closed in 1967, would be re-opened. This would permit an LRT network with three routes running through Piccadilly Gardens in Manchester city centre:

Altrincham – Glossop/Hadfield
Bury – Marple/Rose Hill
Rochdale – East Didsbury via Oldham and Chorlton.

A Bill was submitted to Parliament for powers to construct the cross-Manchester links and to seek government funding for the first route, which was to be between Bury and Altrincham. Here was the genesis of what would become known as Metrolink. Meanwhile background work was continuing on the Windsor Link in Salford which would allow BR services from Bolton to serve Piccadilly – this had first been mooted in 1980 – and on the Hazel Grove chord to allow trains to Sheffield to run via Stockport.

Above **Atlantean deliveries came to an end in 1984. New Atlantean 8738 was painted to promote interest in Manchester's heritage of Victorian architecture. Two qualifying buildings appear in this view – the Cathedral and, on the right, the Corn Exchange. Overall adverts continued to contribute to GMT's revenue.** R L Wilson

Left **A brochure was produced to promote the planned light rapid transit network, showing orange-liveried articulated trams in a traffic-free Market Street. When light rail did come the trams were grey and Market Street was congested with deregulated buses.** GMPTE

A new low-cost (£86,000) station was opened at Humphrey Park, Stretford in October. At the same time the PTE's involvement in the local rail network achieved some highly visible recognition with the repainting of a Class 303 electric multiple unit (emu) in orange and brown. The first to be so treated was 303 060. The Class 303 emus were introduced to the Manchester to Hadfield and Glossop line in December; 12 were transferred from British Rail's Scottish Region where they had been used on Glasgow suburban services since 1961. They replaced 30-year-old Class 506s. The 506s were 1500v DC emus; their demise (probably thought long overdue by commuters on the line) was brought about by the conversion of the route to 25kV operation.

A long line of Leyland Atlantean deliveries came to an end in July with the delivery of 8765 to Oldham. A total of 1,225 standard Atlanteans had been delivered to Selnec and GMT over a 13 year period and when the last entered service standard Atlanteans accounted for around 50 per cent of the GMT fleet. If the 500 standard Fleetlines delivered to Selnec and GMT plus the 90 ordered for LUT were added, no fewer than 1,815 standards had entered service on first-generation rear-engined chassis. Add the 21 prototypes and the total reached 1,836.

1984 was the watershed year for first-generation standards. The figure of 1,815 operational vehicles of the type was never quite reached. Two buses, 7422 and 7656, had been withdrawn in 1983 after fire damage, which left 1,813. But only one month after delivery of the last of the first-generation standards, the first routine withdrawal of a standard took place when 1972 two-door Fleetline 7280 came out of service.

The Atlantean had been phased out of production by Leyland because it did not meet EEC legislation on noise and brake performance. The final 60 chassis of GMT's last Atlantean order – which would have taken fleet numbers for standards up to 8825 – were cancelled and replaced by 60 Olympians. At the same time a new order was placed for 220 Olympians. Standard Atlantean 8620 was used as part of a trial of an improved heating system for the second batch of Olympians which incorporated a heat exchanger under the staircase; coincidentally it later received an Olympian-style front grille panel.

Delivery of the initial order for 25 Olympians, started at the end of 1982, was completed by March. The Olympian featured air suspension and Leyland's automatic Hydracyclic gearbox with integral retarder, but in standard GMT form it offered fewer seats – 73 (43 up, 30 down) compared with 75 (43/32) on Atlanteans and Fleetlines. The Olympians weighed 9877kg, compared with 9550kg for an Atlantean with Northern Counties body or 9400kg for a Park Royal-bodied Atlantean.

For comparison with the Hydracyclic, whose reputation was generally poor because of problems with oil seals failing and hoses breaking, GMT ordered ten Olympians with Voith gearboxes. These were delivered in July and August as 3026-35 and had the high driving position. They were divided equally between Princess Road and Hyde Road, neither of which was already running Olympians.

With the delivery of 3035 the allocation of Olympians in neat groups of five came to an end and subsequent deliveries were allocated to garages in the same haphazard way as standard Atlanteans and Fleetlines had been. During 1984, 67 Olympians joined the fleet and the type had reached all GMT depots except Rochdale and Weaste. Olympian 3036, delivered in Bus & Coach Council promotional livery, spent a week running for Blackpool Borough Transport in September during the period of the BCC's annual conference.

Dennis received its first major order from GMT in 1984. This called for 50 chassis – 30 Dominator double-deck and 20 of a new rear-engined midibus to carry 40 passengers, which was to emerge as the Domino. All were to be bodied by Northern Counties. The midibuses were conceived as standee vehicles with 20 seats and room for the same number of standing passengers. They were to replace the hard-worked Seddon Pennines on the Centreline service in Manchester and were specified with Perkins 6.354 engines, Maxwell four-speed automatic gearboxes and ZF power steering. The first of the Dominos was completed – almost – in time for the Motor Show in October. In actual fact the body and chassis were complete – but the chassis lacked an engine and gearbox as inquisitive visitors could find if they cared to kneel down and peer underneath. The air suspended chassis was expensive and over-engineered, using, for example, Dominator axle beams.

The Domino was a stylish little bus, 25ft 2in long and 7ft 6in wide. The body had a welded tubular steel frame and square-cornered side windows which were bonded to the structure rather than being held in place by rubber gaskets. Bonded glazing gave a crisper appearance but was more expensive to replace than windows held in place by rubber gaskets because of the time it took for the adhesive to cure, which meant keeping a damaged bus out of service for a longer period. A new Centreline livery was devised with white for the main side panels, orange for the roof, and a brown and orange striped skirt. The show bus, 1751, also featured orange handrails and seat frames, although subsequent bodies used conventional polished steel for the internal grabrails. The destination display was set behind the windscreen. There were 24 seats and a large luggage pen over the nearside front wheelarch. The other wheelarches had inward facing seats (for three passengers each) while the forward-facing seats were arranged in two-plus-one fashion to give space for up to 14 standees, giving a maximum load of 38. Attractive as they were, the GMT Northern Counties-bodied Dominos remained a unique combination. The only other Domino chassis built were supplied to the South Yorkshire PTE.

The Dennis order was announced as GMT was receiving three of the rare Dennis Falcon V model. These were 1471-73 and had the highest seating capacity – 84 – of any standard-style buses in the fleet. (The GMT buses with the greatest number of seats were the Bolton-ordered Atlanteans 6802-16, which seated 86.) The 10.5m-long Falcon V featured a compact 11-litre Mercedes-Benz OM421 V6 engine. This was rated at 188bhp and was mounted at the rear of the air-suspended chassis, driving straight to the GKN rear axle via a Voith D851 three-speed automatic gearbox with integral retarder. The simple driveline eliminated the need for an angle drive but gave the Falcon a long rear overhang – 9ft 1in compared with 7ft 8in on an Olympian. The aim of Dennis's designers was to cut weight and cost by ten per cent, while increasing carrying capacity by ten per cent. However the V6 engine was not particularly smooth running and only six Falcon Vs were built. GMT's Falcons arrived in April and were allocated to the former LUT depot at Atherton.

Among GMT's biggest buses in terms of seating capacity were the three Dennis Falcon Vs delivered in 1984. They were 10.5m long and their Northern Counties bodies seated 84. The Falcon re-introduced Mercedes power to the fleet – it had last been seen on the two O.305 single-deckers. The Falcons – which tipped the scales at 9900kg – operated from Atherton. Stewart J Brown

One area of the highly-standardised Leyland National which was subject to change was the design of the roof-mounted heating pod. Later vehicles had higher pods than earlier ones, as Wigan depot was finding out in encounters with a low bridge at Gathurst on the west of the town on route 620 (Orrell Post to Bradley) which carried the railway to Southport. There was also a tight clearance on the 333 to Wrightington Hospital, which passed under the same line at Appley Bridge. The problem was overcome by switching the pods from older 13xx-series Nationals to the 11 ex-LUT examples operating in Wigan.

Deliveries to Charterplan were not quite as planned. The two Setras and four Leyland Tigers with 245bhp Leyland TL11H engines were received, but – in what looked like a replay of the Titan fiasco – Leyland was having severe problems producing its new Royal Tiger Doyen integral coach at the Leeds factory of Charles H Roe. Charterplan was forced to cancel its Doyen and in its place received an ex-Leyland demonstration Tiger, WBV540Y, which became number 25. This had been new in 1982 and had a Plaxton Paramount 3500 body. It joined the Charterplan fleet in March. The two 1984 Setras had been delivered in the winter of 1983. The four Tigers had Plaxton Paramount bodies and entered service in February as Warburton's 54/5 and Charterplan 66/7. The first pair were registered A54 and A155KVM – afraid of upsetting delicate sensibilities, the Department of Transport did not issue A-prefixed registrations including the number 55 lest anyone read A55 as ASS.

The Leopard which should have been rebodied as 89 in 1982 appeared in 1984 as 100, with personalised registration 476CEL. It had originally been 82 (HNE642N) before rebodying and in its new form ran briefly and incorrectly as KDB676P (the original registration of Leopard 89 before it was rebodied) before being reregistered 476CEL. Its Duple Dominant IV body was fitted with a centrally-mounted wheelchair lift on the nearside and its seating configuration could be either 34 seats, 22 seats plus four wheelchairs, or 16 seats with six wheelchairs. It was equipped with a toilet compartment.

Top To compensate for the non-delivery of a new Royal Tiger Doyen integral coach, Leyland provided Charterplan with this ex-demonstration Tiger with Plaxton bodywork. J Robinson

Centre For 1984 Charterplan switched the body order on its new coaches to Plaxton for the first time since 1978. Four Leyland Tiger 245s with Paramount 3200 bodies were delivered: two for Charterplan and two for Warburtons. A Charterplan coach leaves Liverpool on National Express service 960 to Caernarfon. R L Wilson

Right The last of the Leyland Leopards taken out of service for rebodying in 1981, did not re-appear with its new body until 1984. It had a Duple Dominant body which had been constructed to allow access for wheelchair passengers by way of a lift in the wheelbase. The main entrance door was also wider than normal, to assist ambulant disabled travellers. R Marshall

Charterplan moved into dedicated National Express operation in 1984 and at the end of the year repainted five-year-old ex-LUT Plaxton-bodied Leopards 47 and 48 in National Express livery which incorporated small Charterplan fleetnames. They were used on the 962 service between Manchester, Liverpool and Caernarfon.

Continuing an association of supplying small buses to the Strathclyde PTE (as Greater Glasgow had become), two of the withdrawn dial-a-ride Bedford CF minibuses, 1746/47, were sent on loan to Glasgow in April. They never returned, being purchased and then immediately sold by SPTE later in the year.

Dual-door buses had been a short-lived phenomenon in Manchester as in most other British cities and by 1984 those that remained were being singled out for early withdrawal. Unlike many other operators of dual-door buses neither Selnec nor GMT tried rebuilding them by blocking up the central exit. The first standard to be withdrawn was dual-door Fleetline 7280, one of a pair which had entered service with Selnec Northern in Rochdale at the end of 1972. It came out of service in August. The prototype dual-door standards, 6250-4 (originally EX17-21), were also withdrawn in 1984.

In an effort to reduce vandalism two Bolton standard Atlanteans, 7562 and 7795, were fitted with perimeter seating at the rear of the upper deck; others were given moulded plastic seats. Another Atlantean to be rebuilt was 8217, reseated to 15 in the lower saloon and repainted in Airport Express livery as a companion for 7638, which had been similarly treated in 1983. 8217 was later converted back to a fully-seated bus, but 7638 was withdrawn after an accident in 1986, still in Airport Express livery.

Radcliffe bus station, rebuilt at a cost of £305,000, was re-opened in April. Plans for a new interchange in Bolton were announced in 1984 with work to begin in 1985.

The last Mancunian-style Atlantean in the fleet, 1187, was withdrawn in May, leaving only a small number of Fleetlines in service with the distinctive Mancunian body. The last six of these stylish buses (2289/90/92/98, 2301/02, all from the final SVR-K batch) came out of service in mid-December, bringing to an end 16 years of operation of one of the most advanced double-deck bus body designs ever seen in Britain. A start was also made on withdrawing the 20 ex-London Transport DMS class Fleetlines purchased for LUT in 1980. To improve driver comfort on older buses no fewer than 1,483 cab heaters were ordered from KL for fitment to Atlanteans and Fleetlines built between 1972 and 1981.

The events of 1984 were taking place against a backdrop of political uncertainty. The Conservative central government was unhappy about the powers of the largely Labour-controlled metropolitan counties and was engineering their break-up. If this happened the Greater Manchester PTA's membership would be made up of representatives from the district councils in the area – with the hint that they would be free to withdraw from the PTA, thus effectively bringing integrated public transport to an end. Bolton, Bury, Stockport and Wigan had all at some time indicated some interest in resuming their own direct bus operations, but with a Conservative government strongly opposed to public ownership of transport they were not going to be given the chance. Privatisation and deregulation were in the offing, with their own much more serious threats to integration.

The 1983 Festive Fares promotion was repeated in 1984, with free travel on bus services throughout Greater Manchester on Christmas Day and Boxing Day. A new freeze on fares rises being operated by GMC (fares had not increased since April 1982), contributed to a one per cent rise in travel on GMT buses. Excess fares for passengers seeking to evade payment were re-introduced in March 1984 and some 9,000 were collected in the 1984/85 financial year. In addition there were 796 prosecutions for fares offences. An under-16 identity card was introduced in June to help staff identify those children eligible for concessionary travel. At the same time the upper age limit for free travel for young children was raised from three to five years. In September a new Zone 5 Saver Ticket was introduced, extending Saver Ticket availability to New Mills and Hayfield.

Manchester's Centreplan, involving the extension of city centre pedestrianisation, was completed in July. In the area of Piccadilly Gardens, access to Oldham Street, Parker Street and part of Mosley Street was limited to buses and taxis.

With the drop in passengers having bottomed out (at least until deregulation), 1984 was a turning point, loadings remaining static at 342m journeys. The bus fleet was trimmed some more, bringing it down to 2,486, while staff numbers were cut to 9,474.

Two ex-LUT Leopards were repainted in National Express livery in 1984 for services to North Wales. These were 1979 coaches with Plaxton Supreme Express bodywork and 48 reverses out of the bus station at Chester. Stewart J Brown

Dual-door double-deckers had been ordered by the PTE in the early days of one-man-operation and withdrawals started in 1984. 7206 illustrates how the GMT fleetname was located further back on two-door buses. Stewart J Brown

1985: The Domino Effect

To increase sales of off-bus tickets – ClipperCards and Saver Tickets – the PTE extended the sales network from 7th February. Previously tickets had only been sold in 26 GMT outlets and 100 rail stations. Now they were also available from 700 Post Offices in Greater Manchester. The increased number of sales points was actively promoted by advertising on Piccadilly Radio, the local commercial station, and – for the first time – on television. Within 12 months around 40 per cent of GMT's pre-purchased tickets were being sold through Post Offices. At the same time the rail zones for the Saver tickets were altered which led to a price reduction in Zone 3. Agreement was reached with the West Yorkshire PTE in the spring to allow elderly and disabled GMT permit holders to travel for half fare on off-peak WYPTE buses up to a maximum charge of 15p on weekdays, and free on Sundays. WYPTE pass holders were allowed to travel for 10p on GMT buses. GMT permit holders were also allowed half-price travel on buses in Lancashire. At the end of March membership of the TeenTravel Club, which offered cut-price ClipperCards, was extended to include 18 and 19 year olds. Fares remained frozen at 1982 levels and ridership rose yet again by one per cent.

Fleet numbers 1441-3 with registrations B443-5TVU had been allotted to the three Volvo Citybuses in the expectation that they would enter service before August – but in the event they were not delivered until 1986 and the B-prefix registrations were surrendered. However the 30 Northern Counties-bodied Dennis Dominators did arrive in 1985 and were numbered 2001-2030, re-using the series of numbers previously allotted to Mancunian-style Fleetlines. They had 75 seats – two more than on the Olympians being delivered at the same time. Delivery started in March and was completed in May. The Dominators were allocated to Rochdale (2001-5, 2021-5), Bolton (2006-10, 2026-30), Northenden (2011-5) and Oldham (2016-20). All of these depots already ran Leyland Olympians, although at Rochdale there was only one.

Olympian deliveries continued, all fitted with Gardner engines, Hydracyclic gearboxes and Northern Counties bodies. A total of 105 entered the fleet in 1985, taking the fleet number series up to 3190. The growing popularity of the Airport Express encouraged GMT to specify a new Olympian (3139, delivered in July) with only 16 seats in the lower saloon. This left space for luggage. At Northenden 3139 entered service in an overall advertising livery appropriate to its use: SAS – Scandinavian Airlines.

Above **GMT was one of only two users of the Dennis Domino midibus, which had a rear-mounted Perkins engine and a Maxwell gearbox. Twenty Dominos replaced the Seddon Pennines on the Manchester Centreline service between the railway stations at Victoria and Piccadilly, where 1752 is loading. The body was by Northern Counties.** Stewart J Brown

The batch of 25 VRTs ordered by North Western spent its entire life at Stockport and survived almost intact until 1985, by which time some vehicles had acquired the post-1980 GMT livery.
Stewart J Brown

The final reminder of Bolton Corporation was the batch of 15 long-wheelbase Atlanteans which were delivered to Selnec in 1971-72. Five of these were transferred to Hyde Road in 1984 to eke out their last days. They were withdrawn in 1985.
M R Keeley

dual-door standards delivered in 1972-73, most had gone by the end of 1985 – along with a fair number of early one-door standards too. Just over 100 standard Atlanteans and Fleetlines new in 1972-73 were taken out of service in 1985. The survivors from the prototype standards, EX1-21, were also withdrawn during the year.

The 25 Bristol VRTs ordered by North Western and delivered to Selnec in 1973 had survived remarkably well with 22 still in operation at Stockport at the start of 1985. But by July all had gone, breaking GMT's last link with North Western. The withdrawal of the VRTs and of ex-LUT RESLs 414-23 brought to an end GMT's operation of Bristols. Also withdrawn during the year was the last bus with Bolton Corporation links, 6810, an Atlantean ordered by Bolton but delivered new to Selnec.

A new depot was opened in Bury in July, the PTE's fifth. (The others were at Bolton, Stockport, Altrincham and Tameside.) Work had started in January 1983 and involved diverting a stream into a culvert. The depot stood on a 5.3 acre site and was designed to hold 100 buses. It became operational on 8th July and was officially opened by the town's mayor on 22nd August. It cost £5 million.

Ashton's rebuilt bus station was officially opened in March – at the same time as work started on a £1.5 million rebuild of Moor Lane bus station in Bolton which involved new shelters and saw-tooth platforms to replace the existing straight-platform layout.

Unannounced additions to the bus fleet were two Leyland Cubs with Reeve Burgess bodies, 1701/2, used for a new Localine service in Wythenshawe, south Manchester. These were equipped with tail lifts for wheelchair passengers. The new services started on 1st July and were designed specifically for people with mobility handicaps. A flat fare of 30p was charged. Leyland Cub PSVs were rare; most were sold for use as school buses.

Having withdrawn the last Mancunians and the first standards in 1984, a start was made on the early standards in 1985, with particular attention being paid to weeding out the two-door buses. Of the 48 production

The first Dennis Domino, 1751, was delivered in September and the type started to appear in service on the Manchester Centreline in November. The low-floor Domino with its wide entrance was a considerable advance on the front-engined Seddons which had served the route remarkably well for over 10 years. However delivery of the Dominos took until January 1986 and the last of the Seddons did not cease buzzing around on the Centreline service until February. Five of the Seddons, 1731-33/38/42, were repainted in Domino-style livery for services in the Altrincham area.

Major service revisions were made in the west area in September, largely involving former LUT operations in Leigh and Atherton. Some 56 routes were affected. One result of the changes was a reduced requirement for single-deckers and this led to the early withdrawal of some Leyland Nationals.

Free shoppers' services to the Tesco supermarket at Sudden were introduced in the Rochdale area in October. These followed the introduction of a similar contract for an Asda store earlier in the year. The Rochdale fleet was fitted with transponders to give buses priority at traffic lights at 12 junctions in the town.

To improve transport for people with temporary or permanent mobility impairment Greater Manchester Council, GMT and Manchester Community Transport teamed up to instigate a dial-a-ride service, marketed as Ring and Ride, for disabled travellers. The service started in September with two Mercedes-Benz 307D vans with Reeve Burgess conversions. The vehicles were not owned or operated by GMT, although they carried a striking version of the fleet's white, orange and brown livery.

The Charterplan coach fleet received four new Leyland Tigers with Plaxton Paramount bodies (1-4) and a new Volvo B10M GL capable of carrying wheelchairs. The Volvo, 5, had a Plaxton Paramount 3200 body which could be run as a 53 seater or as a 29 seater with seven wheelchairs. In either layout it had a toilet compartment fitted. The fashion for re-registering coaches to disguise their ages was taking hold of the industry in 1985 and Charterplan succumbed by buying so-called cherished numbers for eight coaches:

Fleet no.	Old reg no.	New reg no.
21	BJA856Y	UOV721
22	FWH22Y	628ELX
23	FWH23Y	TXY978
24	FWH24Y	FSV578
25	WBV540Y	4195PX
32	A32KBA	OTK802
33	A33KBA	OXK373
4	B471VBA	TPX884

Registrations containing the letter X, which five of those bought by Charterplan did, were comparatively inexpensive because few people have X as one of their initials.

Charterplan secured additional National Express work with a run on the 825 service from Manchester to Birmingham, Heathrow, Gatwick and Eastbourne, and on the 963 with workings to Llandudno and Sheffield. Two second-hand Leyland Tigers were purchased to cope with the increased National Express commitment. These were KGS491/3Y, 1983 models with Plaxton Paramount 3200 53-seat bodies. They had been new to The Londoners, based in south-east London, and were purchased from Arlington, the London Leyland coach dealer. Both entered service in white and soon had National Express fleetnames added. They were numbered 64/5. The two other National Express-liveried coaches, 47/8, were updated by the fitment of Supreme V front ends to their Supreme IV bodies. An MCW Metroliner single-deck demonstrator, ABM399A, was tried on the 960 but MCW was not a strong player in the coach business and Charterplan did not buy any Metroliners. The introduction of National Express's winter timetable in October saw Charterplan running on the 850 (Manchester, Wolverhampton, Birmingham, Luton, London) and 960 (Leeds, Manchester, Liverpool, Caernarfon) services. The coach used on the 850 also worked a trip on the 037, south from London to Bexhill-on-Sea.

Volvo made a fresh appearance in the coach fleet with the delivery of the PTE's first B10M in 1986. This had a Mark II Plaxton Paramount 3200ls body with deep windscreen and low driving position. It was a 53-seater and was the second coach in the fleet with provision to carry wheelchair passengers. R Marshall

Expansion of Charterplan's interest in National Express operations led to the purchase of a pair of two-year-old Plaxton-bodied Leyland Tigers in 1985. They had been new to The Londoners, a Peckham-based company, and were operated by Charterplan in National Express white livery. G R Mills

The Titans received two different versions of the 1980 livery. Stockport's Leyland-powered 4014, seen in Manchester's Piccadilly Gardens, has a deep skirt, while some others were given a shallower one. Stewart J Brown

116

Both the Warburton and Godfrey Abbott names still survived. Six coaches carried Warburton livery (24,27,53-55 and 85) and two retained Godfrey Abbott colours (52 and 88). Orders were placed for five Leyland Tigers with Duple 320 bodies and two Volvo B10Ms with Plaxton bodies for 1986 delivery.

One event in 1985 overshadowed all others, not just in Manchester but throughout Britain: the 1985 Transport Act. This was set to turn a highly-regulated and co-ordinated industry on its head. Regulation was out; deregulation was coming, and the Government was anxious to cut public transport subsidies. The PTE's bus operation was to be established as a limited company, working at arm's length from the PTE. The PTE would thus cease to be a bus operator and would instead assume the role of an overseeing authority, responsible to a new Passenger Transport Authority for the area's public transport services. A new system of service registrations was to replace the old road service licensing system and all new and existing operators who wanted to run buses after what became known in the industry as D-day, 26th October 1986, had to have their service registrations submitted to the Traffic Commissioners by the end of February 1986.

The first reaction from the PTE to this opening up of the market was a proposal that all buses being sold would have their engines and gearboxes removed, or be otherwise rendered unsuitable for use by potential competitors. This was in fact done to a number of withdrawn vehicles – but what happened in the end was quite the reverse.

A more positive reaction led to an examination of the scope for branding routes and the first to be treated was the 68 (Piccadilly to Little Hulton) which in December 1985 was operated by a dedicated fleet of seven Olympians from Frederick Road, all carrying external advertising for the route.

After many years of complaint from GMC about the ageing rolling stock used by British Rail the first new trains since the 1950s started to appear in September. These were 120-seat Class 142 Pacers (costing a cool £332,000 a set) and the first were put into service on the Victoria-Oldham/Rochdale route alongside the ageing diesel multiple units which they were to replace. The fleet of 14 was based at Newton Heath depot. The Workington-built Class 142 used Leyland National body parts but was built to a width which allowed the use of two-plus-three seating. The first 14 were delivered in PTE orange and brown livery with a white band at cantrail level. Two livery layouts were produced by the PTE's Ken Mortimer. BR chose the version with a brown skirt rather than an alternative which featured an upswept brown band at each end of the train. The Class 142s were powered by a 205bhp Leyland TL11H engine with an SCG gearbox on early units and a Voith gearbox on later examples.

A new unmanned rail station was opened in March at Mills Hill between Middleton and Chadderton on the Rochdale to Victoria line at a cost of £100,000. In May another new station was opened at Flowery Field, Hyde, between Guide Bridge and Newton on the Glossop line. These were followed later in the year by three more stations at Smithy Bridge, between Rochdale and Littleborough; Derker (Oldham) and Ryder Brow, near Belle Vue on the Marple line. Smithy Bridge opened in August and was soon attracting 275 passengers a day. Plans for more low-cost stations were announced in 1985 at Godley Mottram Road (to replace Godley Junction), Gatley Hill, Lostock Junction, Crompton Way, Abraham Moss and Smedley. The last two were to replace Woodlands Road on the Bury line. A planned rail link between Manchester city centre and the airport was under discussion but foundered on cost grounds. It was going to cost between £15 million and £18 million. On a rather smaller scale a new Salford Crescent station was also planned, scheduled to open in 1987.

The Parliamentary Bills authorising the first stage of Greater Manchester's light rail network were making smooth progress and tentative consideration was being given to a line to the developing Salford Quays area. In the meantime the government approved the £12 million Windsor Link which would run from Salford University to Castlefield and permit trains from the Bolton line to serve Piccadilly. Construction was expected to be completed by 1988.

Interest in minibuses was spreading from the south of England and GMT indicated that it was to purchase a fleet to test the potential of such vehicles. An order was placed with Dodge at Dunstable – who in the midst of an identity crisis was also known as Renault – for forty S56 chassis, to be bodied by Northern Counties.

GMT's operations director since 1980, Jack Thompson, retired in 1985. He had come to Selnec from Manchester Corporation Transport where he had worked from 1949, latterly as general manager. He was succeeded by Ralph Roberts, who joined GMT from the National Bus Company.

Roberts took over as passenger numbers were actually rising, up from 342 to 346 million. This was achieved despite a slight cut in the fleet which at 2,432 was now the smallest it had been since 1971 – and this despite the absorption of North Western, Wigan Corporation and Lancashire United. Staff numbers were cut by a bit over 300 to 9,144.

GMT introduced dedicated route branding at the end of 1985 on Leyland Olympians operating on the 68 from Manchester to Little Hulton. Further selective route branding followed. Olympian 3100 shows the effect – spoiled just a little by the fact that it has arrived in Victoria bus station on route 57. Stewart J Brown

1986: Deregulation – The End of an Era

The 1985 Transport Act turned the bus industry on its head. The PTEs had been the creation of a 1960s Labour government which saw virtue in the co-ordination and integration of public transport. The 1985 Act was the culmination of free market Conservative party thinking which saw no benefits in integration and great dangers in subsidy. It was also opposed to public ownership of essential services such as power, water – and transport.

So GMT was set for change.

Reaction was understandably angry after the best part of two decades in which the aim had been to raise public transport standards. The chairman of GMC's passenger transport committee – and of the new Passenger Transport Authority created to succeed it when GMC was dissolved on 31st March – was hard-hitting. Councillor Guy Harkin introduced the 1985/86 Annual Report of GMT with strong criticism of the new Transport Act:

"The new PTA, as much as its predecessor, views mobility for all the people of Greater Manchester as an essential ingredient in the quality of life. The contribution of an integrated and efficient system of public passenger transport to this object is indisputable and its role in the economic and social welfare of our county is inarguable. This new Act, however, with its narrow minded concentration on commercial and 'free market' forces threatens to sacrifice the provision of an integrated public transport system in pursuit of economic dogmatism."

Councillor Harkin went on: "Whilst the new Act is faulty in concept the manner of its introduction by the Government is, quite simply, incompetent. Within a space of less than 12 months the Authority and the Executive are expected to dismantle a regulated system of bus operation which has endured for over 50 years."

The policy on what was to be registered by GMT as the new commercial network from October 1986 was delegated to the four area general managers. Overall some 320 services representing 68 per cent of existing GMT mileage was registered. This left 32 per cent to be covered by the PTE in its new role as the co-ordinating body for transport in Greater Manchester, and this it had to do by inviting tenders from operators (including GMT) to provide those services which had not been registered but which the PTE considered socially necessary. In addition around 40 other operators registered 120 commercial routes in the Greater Manchester area. Deregulation opened the door to bus operation for low-cost operators running ancient buses. Vehicle quality in Manchester, as in many other British towns and cities, was about to deteriorate markedly.

Having registered around 68 per cent of its routes for commercial operation from 28th October, GM Buses was hopeful of picking up much of the rest of the network when it was put out to tender by the PTE. The company estimated that the PTE would be inviting tenders for about 13 million bus miles (out of a total of 21 million which had not been registered commercially). Of that 13 million miles, GM Buses' target was to win some 75 per cent, or around 9.75 million.

It was a miscalculation which was to have grave consequences. By D-day, only five million miles had been put out to tender, and of that GM Buses won just 60 per cent, or about three million miles – which was less than a third of its target.

A 16-19 bus pass for young people was introduced in January, for a nine month experimental period (the end of which co-incided with the start of bus deregulation) and allowed a week's unlimited travel for £4. GMT fares rose by around 12 per cent on 18th May, the first increase for four years, as the operation was being prepared to stand on its own, and it was announced that there would be garage closures and over 2,000 redundancies in the workforce. The concessionary fare went up by 2p from 10p to 12p. Rail fares rose by around 15 per cent.

Three depots closed early in the year. The former Salford Corporation depot at Weaste closed on 19th January. Leigh closed on 3rd February; its work was transferred to the former LUT depot at Atherton. Hindley, another one-time LUT depot, closed on 23rd March and its work was distributed between Wigan, Atherton and Swinton.

Above **Northern Counties-bodied Atlantean 8503 heads past Weaste depot after its closure. The bus wears the new GM Buses west area livery with a band of pale green relief between the wheelarches.** J Robinson

Yet all was not gloom and doom. A new identity had to be developed for the free-standing company to be set up by the PTA to take over from GMT. Consultants were called in and under an umbrella holding company, tentatively called Bus North West, proposals were submitted for the operating companies' fleetnames – four were planned covering the north, south, east and west of GMT's territory. A short-list of ten names was considered – Apollo, Champion, Javelin, Lion, Mercury, Meteor, Pulsar, Swift, Tempo and Valiant. However these, even with the addition of the word bus (to give Apollo Bus etc), were a shade too fanciful for GMT's management and were rejected. In-house suggestions included Park for the southern company, Moss for the western one, Moor for the eastern and River for the northern, each with different liveries – tentatively blue, yellow, green and white respectively. Another proposal was to use the word Trans as a prefix to the main towns in the area, in the style of TransWigan, TransBolton and so on.

In the end the company went for a rather more prosaic title and identity. The new operation was to be known as Greater Manchester Buses Ltd with four operating areas – north, south, east and west – based on the existing PTE divisions. The fleetname chosen was GM Buses.

Each of the areas was distinguished by a narrow band of relief colour at wheelarch level on the otherwise unaltered GMT livery of brown, orange and white. There was also a local fleetname displayed above the entrance door. The work was done in haste and the area identities were weak, looking more like a gesture to the staff than a strong marketing move for the passengers. Interestingly, as all this was being planned two buses still survived in LUT livery – Plaxton-bodied Leopard 460 and standard Fleetline 6917.

The decision to retain the GMT livery was prompted by two considerations. Firstly, the travelling public had a positive image of the PTE's bus operation and GM Buses' management was anxious to emphasise continuity in a time of great change. And there was also the question of costs. Even if a new livery were desirable, the fledgling company might have more pressing considerations than repainting its fleet.

The East area (with its headquarters in Oldham) used yellow relief and the local fleetnames City East (operations from Hyde Road), Glossop, Oldham and Tameside. For North (headquartered at the Frederick Road offices) the relief was fawn and the fleetnames were City North (Queens Road), Bury and Rochdale. South, based in Stockport, used grey relief with City South (Princess Road), Trafford (for Altrincham) and Stockport as its local identities. West, operating from the former LUT headquarters at Atherton, had pale green relief and the local names Bolton, Salford (for Swinton depot), Wigan and Atherleigh (for Atherton), a contrived combination of Atherton and Leigh. The use of town names was designed to encourage brand loyalty in a competitive market, but carefully avoided the word Manchester – although it was implicit in the GM Buses fleetname. The relief colours started to be applied to vehicles from mid-July. The M-blcm was retained by the PTE.

The long term aim was to establish the four areas as individual companies in the style of Greater Manchester Buses (East) Ltd etc, each with its own managing director. But this was not implemented as GM Buses reeled under the onslaught of competition.

Withdrawn standard 7173 had been repainted in an experimental livery at the end of 1985 as a pilot for a new identity for the fleet's double-deck express operations. It was even fitted with shrouds over the engine bay to make it look more like the new generation of Northern Counties standard bodies on Leyland Olympians, which did not have the protruding engine compartment of the Atlanteans and Fleetlines. The offside of 7173 was a pale orange (sometimes described as coral); the nearside white. Relief was provided by brown, red and orange stripes.

Park Royal-bodied Fleetline 7173 was painted in a trial for the predominantly pale orange livery being developed for express services. The body was modified by the addition of shrouds above the engine compartment, to make it look more like the new generation of Olympians and Metrobuses which would wear the new colours. It is seen in Charles Street, Stockport; it was not used in service in this livery.
J Robinson

From 26th May 1986 the new express livery – pale orange – appeared on six repainted Tameside Olympian buses running on the Trans-Lancs express from Bolton to Stockport, service 400. Some had their bus seats retrimmed in the red moquette which was to be used on new coach-seated double-deckers.

Deliveries of standard Olympians continued, but from 3239 changed from 73-seat buses to 69-seat coaches (43 up, 26 down) in the pale orange Express livery. The 38 coach Olympians, 3239-77, were allocated to Altrincham, Atherton, Bolton, Bury, Oldham, Tameside and Wigan. The first dozen or so were delivered without seats and stored at Bolton, awaiting the arrival of coach seats from Northern Counties. The coaches were just over 250kg heavier than the Olympian buses, at 10130kg. As the new coach-liveried vehicles entered service, so the new Express image appeared on an increasing number of limited-stop routes.

Registrations for new Olympians were in fact booked up to 3300 (D300JVR) but in the end deliveries finished at 3277 which left 28 buses to be delivered. GM Buses, uncertain about its future financial position, was reluctant to take on a further 28 new buses and efforts were made by the PTE working with Northern Counties and Kirkby Central, the coach dealer, to find a buyer. In the end, a requirement by London Buses for vehicles for its new Bexleybus operation resulted in them being leased by Kirkby Central to London Buses. They were delivered to London at the end of 1987 as E901-928KYR. A replacement batch of chassis and bodies was supplied to GM Buses in 1988 as 3278-3305.

Following the introduction of Olympians 3139 and 3198 on the 200 Airport Express service from central Manchester in SAS and Singapore Airlines livery, two similar buses, 3213/4, were reseated and converted to carry luggage in the lower saloon and all four were painted in a new variant of the orange and two-tone blue Airport Express livery. They differed from earlier double-deck Airport Express vehicles in having coach seats; 3139/98 were delivered as coaches while the other two were converted from buses.

A further 30 Metrobuses, delivery of which started in July and extended through to February 1987, had 72 coach seats (43 up, 29 down) and were the first – and only – Metrobuses to be bodied by Northern Counties. All wore the pale orange Express livery. The first ten, 5201-10, were DR132/8 models and had Cummins L10 engines, while the last 20, 5301-20, were type DR102/51 and had Gardner engines. Cummins had been trying hard to establish itself as a supplier of bus engines in the UK and 5201-10 had the first in the GMT fleet. The ten Cummins-powered buses were divided equally between Tameside and Oldham while the 20 with Gardner engines went to Rochdale (12), Hyde Road (three), Stockport (two) and Oldham (three). The decision to specify Northern Counties bodies reflected the PTE's ownership of the Wigan coachbuilder and its desire to keep the factory busy at a time when many operators were reducing their orders for new buses because of uncertainty over the effects of deregulation.

The unique combination of MCW Metrobus underframe and Northern Counties body was chosen by GMT for 30 coach-seated double-deckers delivered in 1986-87. These were in the express livery. Cummins-powered 5202 pulls out of Bolton bus station on the Trans-Lancs Express to Stockport. Stewart J Brown

A trio of Volvo Citybuses introduced mid-engined double-deckers to Greater Manchester. The Citybuses had 79-seat Northern Counties bodies and were allocated to Wigan. 1483 is seen after the formation of GM Buses. Stewart J Brown

The other double-deckers delivered in 1986 were the fleet's first Volvo Citybuses, 1481-3, which had 79-seat Northern Counties bodies (46 up, 33 down) and were allocated to Wigan which also housed 1446-8, the Volvo-powered Ailsas. The Citybus was derived from Volvo's B10M coach chassis and had a mid-mounted horizontal Volvo 9.6-litre turbocharged engine.

Four examples of Leyland's new Lynx single-decker were delivered to GMT, its first full-size single-deckers since 1977 (or 1979 if LUT is included). These started a new fleet number series, becoming 501-4. They had 48-seat single-doorway bodies built by Leyland and two each were allocated to Rochdale and Bolton; Rochdale's moved to Oldham in 1988. They had Gardner turbocharged 6HLXCT engines and ZF automatic gearboxes and were the only Gardner-powered single-deckers in the fleet. The Lynx was Leyland's replacement for the National and was built at Workington; these were to be the last new Leyland single-deckers purchased by GMT or its successor, and its only Lynxes. The choice of a ZF gearbox reflected widespread dissatisfaction with the standard Leyland Hydracyclic unit.

One of the Dennis Dominos due at the end of 1985 was retained by Dennis for use as a demonstrator. This bus, 1760, was painted in London Buses red livery with white roof, numbered DMB1 by London and used on the C11 service running between Archway and Brent Cross, operating from Holloway garage for a few months at the start of 1986 before being delivered to Manchester. The 19 GMT-liveried Dominos were in operation by March and were allocated to Princess Road (three), Queens Road (six), Hyde Road (five) and Frederick Road (five).

A rather different single-deck addition to the fleet was former LUT standard Fleetline 6938 which re-appeared at Bury in April, rebuilt by Northern Counties as a 28-seat single-decker after an accident at Wargrave Bridge in 1985. It was numbered 1697, to match its registration TWH697T, and was in standard GMT livery with a brown skirt and white roof. In this guise it gave GMT and GM Buses a further five years service. Before deciding to have it rebuilt GMT's engineers inspected a similar conversion of an Alexander-bodied Atlantean by Strathclyde PTE. This bus visited Bury but was not used in service.

Top **Four Lynxes were delivered to GMT in 1986. The first is seen in the centre of Oldham.**
Stewart J Brown

Centre **London Buses expressed an interest in the Dennis Domino and 1760 was painted red and used in service in north London for a few months in 1986 before delivery to GMT. At this time nobody could have foreseen the role Dennis midibuses would play in the capital's transport.**
Stewart J Brown

Right **Five years after the withdrawal of the last single-deck Fleetline, another was introduced to the fleet. But this one was not a purpose-built SRG model; it was instead a chopped-down CRG double-decker. It had originally been part of the LUT fleet. In its new guise it operated in Bury.**
Stewart J Brown

121

Charterplan received seven new coaches in the spring. Two were Plaxton-bodied Volvo B10Ms (6,7) capable of carrying wheelchairs; the other five were Leyland Tigers with Duple's short-lived 320 bodywork. Displaced Leopards 82/6 were repainted in GMT bus livery, as were former Airport Express coaches 35/7. Charterplan continued to provide coaches for National Express and Tiger 23 received a repaint in National Express white in July, followed by 24 and 52 in the autumn. Tigers 53-55 retained Warburton livery.

Competition appeared on Manchester's streets in advance of deregulation with the introduction in April of Citibus Tours service 62 from Blackley to the city centre using one-time Preston Corporation Leyland Panthers in a smart blue and black livery. Citibus also briefly ran a city-centre circle service in Manchester which, with a 40-minute frequency, was perhaps doomed to failure before it started. Finglands, up till 1986 a coach operator, introduced services to the University and UMIST. And old-established independent Mayne, which had co-existed with Selnec and GMT (and Manchester City Transport before them) revised its services in June with routes running out as far as Oldham, Mossley, Glossop and Uppermill.

The spread of minibus operation amongst National Bus Company subsidiaries coupled with the approach of deregulation made a number of major urban operators consider the role of minibuses in their operations. GMT had at the end of 1985 placed an order for 40 Dodge S56s to be bodied by Northern Counties, the first minibuses to be bought by GMT other than for the Sale dial-a-ride fleet. These were 18 seaters, built on the standard Dodge chassis/scuttle with a short bonnet. They had square-cornered glazing and wide doorways with four-leaf double-jack-knife doors, features which were abandoned on subsequent orders. Delivery of the 40 Dodges, in new number series 1801-40, was completed in July. They were shared between Tameside (15), Altrincham (14), Wigan (four), Swinton (four), Rochdale (two) and Bury (one). They entered service in standard GMT livery with the first running in Ashton on service 337 from Hazelhurst to Crowhill, which was marketed as Ashton-mini-lyne. In a similar vein Rochdale got Weaver services, while Chaser routes appeared in Cheetham Hill, Langley, Moston, Whitfield and Hopwood. Hopper minibus services were started in Altrincham, but Wigan's had to make do without catchy names. The names were all short-lived. With the advent of GM Buses the standard GMT livery was quickly abandoned on minibuses, which were re-branded from February 1987 under the Little Gem name in a bright white, brown and grey livery.

Withdrawals in the early part of the year included the last of GMT's dual-door buses, standards 7206/7, which came out of service in March. Massive cuts in the fleet came at the end of the summer. At deregulation, 450 buses were withdrawn as the fleet was trimmed to meet the perceived needs of the new competitive world. In a move which eliminated the Daimler CRG6LXB chassis from the operational fleet overnight the mass withdrawals included all 235 standard Fleetlines remaining from the 7151-7500 batch. Also withdrawn were the first 11 ex-LUT FE30AGR Fleetlines, 6901-11, which were only 10 years old. With them went all of the surviving semi-automatic standard AN68 Atlanteans apart from open-toppers 7032/77. There were 129 in all, but none were newer than N-registered 1974 models. This meant that the lowest-numbered standard (apart from the open-toppers) was 7560.

Over 70 Leyland Nationals were taken out of stock, including R- and S-registered examples. And a few oddities were disposed of – the 15 Leyland Titans, 4001-15, which were on average only seven years old; the ten Metrobuses with hydraulic brakes, 5001-10, which were the same age; and the four original Dennis Dominators, 1437-40.

Because the new GM Buses relief colours had been applied to vehicles at random a number of re-liveried vehicles were amongst those taken out of service at deregulation and did not actually operate for the new company.

It was a massive fleet reduction which merely reflected the cuts being made to services, to depots and to staffing levels. Whatever inefficiency there may have been in major public sector bus operators, they were paying dearly for it now. Integration, still the transport aim in mainland Europe, had turned into disintegration in Britain – and Manchester's bus users suffered badly.

Three more depots closed on D-day, 26th October. Although the closures were part of GMT's plan, the politics involved in getting ministerial approval meant that formal go-ahead for the closures was given only three days before D-day – it was little wonder that deregulation brought chaos to the streets of Greater Manchester. The depots involved were the ex-Manchester Corporation sites at Northenden and Birchfields Road, and Salford's Frederick Road. The portion of their work which remained after D-day was transferred to Stockport, Princess Road, Queens Road, Hyde Road and Swinton. The three closed depots were then used to house the 450-plus withdrawn buses which had been taken out of service as a result of deregulation, and ownership of which had been retained by the PTE. The PTE set up a joint venture company with coach dealer Kirkby Central to dispose of the unwanted vehicles – a fair number of which found their way back onto the streets of Greater Manchester with small operators.

In addition to the vehicles retained for disposal by the PTE, GMT kept a number of vehicles in a reserve fleet for possible use against new competitors.

At deregulation on 26th October, GM Buses took over 80 per cent of the GMT fleet. GM Buses started life with 2,003 buses and coaches:

Double-deck
Standard Atlantean	1,015
Standard Fleetline	235
Leyland Olympian	268
MCW Metrobus	208
Dennis Dominator	30
Dennis Falcon	3
Ailsa	3
Volvo Citybus	3
Scania	2
Leyland Titan PD2	1
Total	1,768

Single-deck
Leyland National	66
Dodge S56	50
Leyland Leopard bus	30
Leyland Tiger coach	22
Dennis Domino	20
Seddon Pennine midi	17
Leyland Leopard coach	16
Leyland Lynx	4
Setra coach	4
Volvo B10M coach	3
Leyland Cub	2
Daimler Fleetline	1
Total	235

TOTAL FLEET 2,003

Both sides of a leaflet produced in the run-up to deregulation for each of the four new divisions.

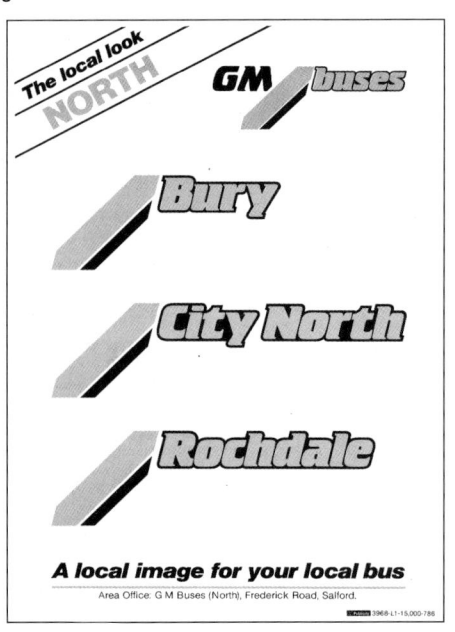

GM Buses started with 5,899 employees – just under half the number employed by GMT only 10 years earlier.

The PTE was awarding tenders to cover gaps in the network registered by GM Buses and other operators – such was the volume of work that some tenders were not awarded until the eve of deregulation. The net result was chaos – and a sharp drop in bus use. After modest increases in bus patronage in 1985 and 1986 which brought the annual passenger figure up to 350 million from a low of 342 million only three years earlier, there was a 10 per cent drop after deregulation. Indeed the problems, especially in the early post-deregulation days, made a nonsense of GM Buses' advertising which featured on posters and television with the theme: "Pick you up tomorrow as usual". Which is precisely what many of its services singularly failed to do. Hardest hit were the North and South areas where the network had been extensively revised. In the East and West the pre-deregulation routes had been preserved as far as possible, with less disruption to travellers.

Bad news for the buses was good news for rail. One of the beneficiaries of deregulation in Greater Manchester was British Rail, where patronage went up by four per cent. Earlier in the year – in May – rail services in south Manchester had been extensively revised as a result of the opening of the Hazel Grove chord which saw new BR Trans-Pennine services being introduced with connections from the Sheffield line via Stockport to Piccadilly. The first of BR's Sprinter units for the Manchester area arrived in March.

A start was made on painting the 17 Bury line Class 504 emus in PTE livery – two were treated by early summer – similar to some of the Longsight-based Class 303s, of which 303 060 and 303 067 were in orange and brown, with more soon following. The PTE-liveried Class 142 Pacers, of which there were 14, tended to be used on services to New Mills and Rose Hill. On the Oldham/Rochdale loop line two-tone blue provincial-liveried examples were the norm. These were generally replacing 30-year-old Class 104 dmus.

Bolton's Moor Lane bus station reconstruction was finished in the autumn and work started on a new bus/rail interchange at Trinity Street in Bolton and a new bus station in Wigan. A new rail station was opened at Hall i'th'Wood, between Bolton and Bromley Cross, and at Mottram Road, Godley where the old station was renamed Godley East.

An interesting expansionist move at the time of deregulation saw GM Buses establish services in Lancashire, operating under contract to the County Council. GM Buses won tenders to run local services in Blackburn, along with off-peak and Sunday services which took GM Buses vehicles as far away as Preston and saw Olympian coaches running between Chorley and Blackburn. The foray into Lancashire was not a success. The operations were run from Bury and were remote from the company's main business. GM Buses received warnings from Lancashire County Council over service reliability. It pulled out of its Lancashire contracts in January 1987 when the settling-in period for bus deregulation was over.

The first 40 minibuses for GMT were Dodge S56s with Northern Counties bodies with wide doors and square-cornered glazing. Both features were abandoned on subsequent deliveries. 1826 is seen in the original short-lived GM Buses minibus livery with its local identity – Wigan – above the entrance. Stewart J Brown

The Hopper name with a rabbit logo was used briefly at Altrincham, as seen on white-liveried Dodges in Altrincham depot. These buses had a single-width entrance and conventional gasket glazing. The former increased seating capacity; the latter cut costs. Altrincham-based buses carried Trafford as their local fleetname. J Robinson

Following the success of the Manchester Ring-and-Ride service for disabled travellers, a similar scheme was launched in Wigan in October 1986. Others, partly funded by the PTE, soon followed in other towns in the region.

The major changes which resulted from deregulation did see operating costs per mile at GM Buses fall by almost 25 per cent in 12 months. At the same time, bus miles per employee rose 23 per cent. The operation was certainly becoming more efficient.

GM Buses: the initial fleet

Coach	1-12, 21-25, 27, 30-33, 35, 36-38, 47/8, 52-55, 64-67, 81/2, 85/6, 88, 90/1, 97, 100
Leyland National	132-4/6-41/4/5/7/8, 150/1/3/4/7/9-62/5-7/9, 170/6-80/6-95/7-203/5, 228, 232/4-40/3/5-50
Leyland Leopard	435-464
Leyland Lynx	501-504
Ailsa	1446-48
Leyland Olympian	1451
Scania	1461-62
Dennis Falcon	1471-73
Volvo Citybus	1481-83
Fleetline single-deck	1697
Leyland Cub	1701-2
Seddon midi	1714/6/7, 1720-24, 1730-34/8-40/2
Dennis Domino	1751-70
Dodge S56	1801-50
Dennis Dominator	2001-30
Leyland Olympian	3001-3267
Leyland Titan PD2	3270
MCW Metrobus	5011-5190, 5201-10, 5301-18
Standard	6912-37/39-90, 7032/77, 7208/15/6/23/7/32/46/81, 7560-7638, 7640-55/7-7960, 8001-26/8-8172/4-8765

A few vehicles survived in the original Selnec-style livery in 1986. Metrobus 5068 carries its new GM buses identity, complete with the short-lived relief colour between the wheelarches, while Atlantean 7142, soon to be withdrawn, remains in white and orange. *Stewart J Brown*

Two faces of the Airport Express. Standard Atlantean 7638 carries a blue relief band between decks to promote the service, while Olympian 3139 has an overall advert for SAS. They are seen approaching Piccadilly station in the summer of 1986. *Stewart J Brown*

The new-generation Leyland Titans had a short life with GMT and all were withdrawn in the autumn of 1986 as part of a major fleet reduction. A few months before its demise, 4007 arrives in Manchester from Glossop. *Stewart J Brown*

The two Leyland Cubs were based at Northenden and operated in Wythenshawe where there were two Localine routes – 1702 is arriving at Wythenshawe bus station on route B. Reeve Burgess built the bodywork. *Stewart J Brown*

125

The Late 1980s: A Competitive Market

After the initial shock of deregulation – and the loss of around £40 million worth of subsidised income – GM Buses eventually settled down in the new competitive environment.

Having been disappointed by the relatively low route mileage put out to tender by the PTE in the run-up to deregulation, GM Buses hit further difficulties in this area over the following 12 months. During this period the PTE awarded over 350 tenders covering some 4.5 million miles – but GM Buses had neither the staff nor vehicles to be able to make a bid for this business.

This left the field wide open for new operators – which helps explain why Manchester soon became one of the most hotly-contested areas of urban bus operation in Britain.

Before settling down though, the area structure was quickly changed and in April 1987 the North area was disbanded. This cut the company's overheads without reducing services. Queens Road and Rochdale depots moved to the East area; Bury went to the West. At the same time Hyde Road moved from East to South and its buses were rebranded City Central instead of City East. In the run-up to deregulation the new area management had been encouraged to foster a sense of area identity among the staff. Some naturally felt that such a quick change devalued the hard work put into building up the areas.

Amidst growing concern about the risk of attacks on passengers and staff, a new feature to appear in GM Buses' vehicles was the on-board video camera. As well as deterring physical attacks, the cameras were designed to cut vandalism and graffiti too. The first were installed in 1987.

Competitors came and went. Some were small, like the Lyntown Bus Company with its smartly-liveried Bristol REs. Some were large, like United Transport's Bee Line Buzz Company with 225 new minibuses competing on routes southwards from Manchester and around Stockport.

Some impression of the scale of minibus competition in 1987 is shown facing page upper with a GM Buses Dodge being followed by three Bee Line Sherpas and a GM Buses Iveco. One of GM Buses' major worries was United Transport, which launched an intensive network of minibus services in Stockport and South Manchester as The Bee Line Buzz Company facing page lower. It was a real attempt to raise standards which failed as both GM Buses and other operators responded. This is one of Bee Line's original fleet of Northern Counties-bodied Dodges. Above left An MCW Metrorider waits for custom in central Manchester. Above right Deregulation brought havoc to the streets of Manchester with long queues at enquiry offices as confused passengers – and sometimes equally confused staff – came to terms with the changes. It made the news headlines – even if the writer couldn't spell 'chaos'. Stewart J Brown

United Transport had announced its plans in the autumn of 1986. GM Buses originally wanted to come to some arrangement with United Transport, taking a minority share in its new Manchester company. This would have avoided wasteful head-to-head competition. But United Transport had interests in South Africa, and this was unacceptable to GM Buses' political masters. So instead of co-operating, GM Buses had to fight and it quickly placed further orders for minibuses, with the choice of vehicles being led largely by availability.

Iveco Ford and bodybuilders Robin Hood of Fareham got an order for 50 vehicles (1501-50) while 80 Metroriders (1601-80) were ordered from MCW. Further orders for Dodges with Northern Counties bodies raised the total fleet of this type to 230 out of a minibus fleet numbering 361 when deliveries were completed in 1987. The odd one was an ex-Iveco Ford/Robin Hood demonstrator, D240JTU, which operated on loan as 1240 and was then taken into stock as 1500. The first bus from the second batch of Dodges, 1841, was exhibited by Northern Counties at the 1986 Motor Show, still in standard bus livery. However this was soon changed, and around a dozen Dodges were delivered in white before the new Little Gem livery was settled on and applied to all Dodges from 1867 upwards. The Metroriders and Ivecos carried Little Gem colours from the start.

The Dodges were spread throughout GM Buses' territory but initially the Metroriders were divided between two depots – Altrincham with 60 and Princess Road with 20. The 51 Iveco Fords started life at Stockport (44), Atherton (three) and Rochdale (four) but before long they were moved out of Stockport to give Hyde Road a substantial allocation (28) with the remaining 33 ending up at Rochdale. One Dodge, 1986, spent some time with Fylde Borough Transport followed by shorter periods at Badgerline and Cleveland Transit, only arriving in Manchester in February 1988 – four months after the rest of the batch. A further two, 1847 and 1950, ran for Fylde in August 1988, with their "I'm a Little Gem" branding altered to "I'm a Baby Blue", which was Fylde's marketing name for minibuses.

GM Buses anticipated an eight year life for the Metroriders with a mid-life refurbishment after four. Other types were expected to run for four or five years. The company did not buy any of the lightest minibus types – the Ford Transit and Freight Rover Sherpa – but it did hire five new Sherpas from Ribble in November 1986 for operation in Wigan. These, D567/9/70/3/4VBV, had 16-seat Dormobile bodies and operated in white with GM Buses fleetnames.

Bee Line, whose activities played a major part in GM Buses' interest in minibuses, is almost a case history of the volatility of deregulated urban transport. It started running in January 1987. By the spring of 1988 it was clear that it wasn't the success United Transport had expected. GM Buses was interested in buying the company in its entirety, or in taking a shareholding. This would have given GM Buses a low-cost operating unit which would have allowed it to expand outside Greater Manchester with more success than its abortive Blackburn area sortie in 1986. But it proved impossible to reach agreement and in September 1988 Bee Line was sold to Ribble.

In April 1989 Ribble was bought by Stagecoach, and in September Stagecoach sold the Bee Line operation to Drawlane (which later became British Bus) who quickly abandoned United Transport's aim of high-quality high-frequency services and expanded the Bee Line fleet with time-expired double-deckers.

Other early competitors for GM Buses included Wigan-based coach operator Shearings, East Midland Motor Services, Ribble, North Western and a very large number of smaller companies. Indeed the size of the United Transport threat diverted some of GM Buses' attention away from other areas. Although United was flooding the South area with a large number of small buses, if competition was measured in terms of seat-miles being offered, other areas, such as the West, had much more to cope with.

At the 1987 Bus and Coach show, Northern Counties exhibited another small bus in GM Buses livery, this time on an Iveco Ford chassis. This was not taken into GM Buses' fleet – but nine Iveco Fords which had been built as stock vehicles by Northern Counties did run on hire to GM Buses in the last six months of 1988. They operated from Oldham in white with Little Gem logos and temporary fleet numbers E181-9, to match their registrations, E181-9CNE.

New bus purchases by GM Buses slowed down dramatically after the influx of 361 minibuses in 12 months. Indeed, second-hand double-deckers made a rare appearance in the fleet in 1988. The last had been ex-London Fleetlines for LUT; the latest were ex-London Routemasters for operation in Manchester. Repainted in London red and running with Piccadilly Line fleetnames, ten Routemasters numbered 2200-09 reintroduced crew operation to Princess Road depot on service 143 from Piccadilly Gardens to West Didsbury from 5th September 1988. They ran until June 1990, when the service reverted to conventional rear-engined buses, initially running in GM Buses livery but retaining the Piccadilly Line brand name. Services 46 and 47 were for a while promoted as the Victoria Line.

There were other branded routes too, as the company explored new marketing ploys to counter competition. In the autumn of 1987 Olympians on the one-time LUT service 582 carried advertising proclaiming "Original 582 service" to combat a copycat service being run by Ribble. Atlanteans on route 50 between central Manchester and East Didsbury – "Over 50 years, over 50 times a day" – also received promotional advertising, as did the three Wigan Citybuses, for the 627 to Skelmersdale.

Top In 1988 GM Buses joined the growing number of provincial operators who bought ex-London Routemasters as part of their post-deregulation competitive strategy. Ten were operated on the 143 from Manchester Piccadilly Gardens to West Didsbury. They were repainted in London red and carried Piccadilly Line branding. J Robinson

Centre At the 1987 Bus & Coach show in the National Exhibition Centre, Northern Counties exhibited an Iveco in Little Gem livery. It was not taken into stock by GM Buses but nine vehicles of this type were operated on hire in overall white. Stewart J Brown

Left In the winter of 1986-87 GM Buses operated services in Lancashire. These were mainly evening and weekend routes which were operated commercially at other times by Ribble and had been put out to tender by Lancashire County Council. An Atlantean waits for customers in Blackburn bus station, carrying GM Buses' somewhat inappropriate deregulation advertising. Many services failed to meet the advert's promise. Stewart J Brown

Route 50 was one of a small number of services chosen for special promotional advertising, as seen on Atlantean 8661. Stewart J Brown

To combat competition from Ribble, buses on the 582 from Bolton to Leigh carried route branding which proclaimed the GM Buses service as "the original 582". Olympian 3237 loads in Bolton in 1988 with a Ribble imposter – ironically a former GMT Atlantean – in the background. Stewart J Brown

The last of the Seddon midis were withdrawn in 1987. 1721 loads in Bury Interchange. Note the Chaser name, which was briefly carried by a few Dodge minibuses before the adoption of Little Gem as a company-wide branding. Stewart J Brown

The success of the Little Gem minibuses saw a short-lived attempt to promote double-deckers on two minibus replacement routes as Super Gems. The routes, 571/2 in Bolton, were run by five Olympians which carried Super Gem names from September 1988. But the Super Gem branding was not a success and was quietly abandoned.

Rebuilds of note in the late 1980s included standard Atlantean 8172, converted to open-top in 1987, and Olympians 3213/4 which were reliveried for the Airport Express and fitted with 61 coach seats (43/18 plus luggage) in 1988. Atlantean 8217 came off the airport service and reverted to bus seating and standard GM Buses livery. Leyland National 249 was adapted in 1989 to carry wheelchairs, being downseated to 24 passengers and fitted with a centre door and wheelchair lift. It had space for up to five wheelchairs and was for use on Localine services in the Wythenshawe and Stockport areas – the routes originally launched with the two Reeve Burgess-bodied Leyland Cubs.

At the end of 1988 the GM Buses fleet totalled 2,025 buses comprising:

Double-deckers	1,568
Single-deckers	52
Midibuses	20
Minibuses	355
Coaches	30

These were split between the company's depots as follows:

Altrincham	99
Atherton	110
Bolton	159
Bury	116
Hyde Road	184
Oldham	130
Princess Road	231
Queens Road	192
Rochdale	111
Stockport	229
Swinton	139
Tameside	132
Glossop	16
Wigan	149
Charterplan	28

Some of the older coaches operated by GM Buses ran in fleet livery rather than as part of the Charterplan coach operation. Leyland Leopard 86 was one of those rebodied by Duple in 1981-82. G R Mills

Route branding was generally applied to the fleet's more modern buses, such as this Olympian which is promoting the Whitworth Valley Way from Rochdale to Rawtenstall. Malcolm King

The biggest minibus fleet (85) was at Stockport, followed by Altrincham with 49. Swinton had the biggest allocation of single-deckers – 11.

In an effort to retain customer loyalty in a competitive market, GM Buses launched its own range of pre-purchased tickets, valid only on its services, in February 1987. The new tickets were sold as Busabout and were competitively priced. In 1988 Busabouts cost £7.70 for seven days, £29.15 for 28 days or £307 for a year, each offering virtually unlimited travel on GM Buses' services. Off-peak Busabouts were also on offer at £4.75 for seven days or £8.70 for 14 days. A Sunday Busabout cost only £1.50. GM Buses, whose predecessor in the shape of GMT had had a virtual monopoly of bus services in Greater Manchester, calculated that its market share had dropped to 66.7 per cent (in terms of mileage operated) by 1988.

GM Buses' first new double-deckers were its last Olympians and its last Leylands, 28 with Gardner 6LXB engines and ZF gearboxes built to replace those which had been sold to London Buses in 1987. They were delivered in the winter of 1988-89 as 3278-3305 and had Northern Counties bodies with

A batch of 28 new Olympians in 1988 introduced an updated Northern Counties body with a new front panel and peaked front domes. 3282 shows the new style, while 3272 behind illustrates the previous front-end design. *Stewart J Brown*

a revised front end featuring a new dash panel and peaked front domes. An additional small window was provided to the rear of the entrance to improve nearside visibility for the driver. Inside there was the new orange moquette first seen on minibuses and the four Leyland Lynxes, and "Bus stopping" signs were fitted, reviving a feature last seen on the Mancunians.

The new Olympians introduced a revised livery to replace the 1981 brown, orange and white. It was a simpler scheme using a stronger orange with white relief restricted to the area around the windscreen, the headlamps and the top deck window surrounds and roof. The entrance doors were white too. Single-deckers had orange below the waist and white above. The half-hearted area colour-codes were quietly abandoned – and actually physically removed from most buses which carried them.

The disappearance of the area identities marked another change in the structure of GM Buses. The three areas – East, West and South – had been set up partly with privatisation in mind. Each was a potentially viable stand-alone bus company, and it was the government's stated intention to see the former PTE bus operations move into the private sector. Alongside the three bus companies there were three other autonomous units which could also have been sold. These were GM Property Services (which had won the contract to maintain the PTE-owned bus stations, among other things), Charterplan, and GM Engineering which was based at Charles Street and was doing a lot of maintenance and repair work for other operators.

However the political climate in Greater Manchester, which had never been very warm towards the Conservative central government, was getting cooler still. The three operating areas were disbanded and control was handed over to a new set of garage general managers, which included appointments from outside the bus industry. Now there was not even lip service being paid to privatisation.

The depot identities were retained, with new pictorial logos and the fleetname became GM Buses with the subtitle "People on the move". The only buses delivered in this livery were Olympians 3278-3305 – although 3278 was actually painted in the old orange but was hastily repainted by Northern Counties before being delivered.

The new livery was soon being applied to existing vehicles as they became due for repaint, although it wasn't long before attempts were made to improve it by the addition of a shallow black skirt. This was applied to four different types of vehicles at Tameside – a Park Royal standard Atlantean (7872), a Northern Counties standard Atlantean (8695), a Metrobus (5049) and an Olympian (3156). Atlantean 7872 also had a black GM Buses fleetname in place of orange.

After the Olympians there followed a two year gap during which buses were withdrawn without replacement, cutting the GM Buses fleet from just over 2,000 at deregulation to under 1,700. The withdrawals included buses which had been kept in reserve in case they were needed to fight new competitors.

Two demonstrators were examined in the summer of 1989, Northern Counties-bodied Renault PR100 F100AKB and DAF SB220/Optare Delta F372KBW. Neither entered service. One unusual hired vehicle was tried in service. One of the Stagecoach Group's three-axle Leyland Olympians, Cumberland Motor Services F202FHH, was used by Hyde Road depot on the 192 from Piccadilly to Hazel Grove in September 1989. It had a 96-seat Alexander body. Brief interest was shown in the new Dennis Dart, with Duple-bodied G350GCK being borrowed for a two-day trial on service 369 between Stockport and Manchester Airport in December 1989. No Darts were ordered.

In 1989 GM Buses replaced the Almex ticket machines which it inherited from GMT with new Wayfarer machines which provided more detailed statistical analysis.

As the competition eased slightly – and particularly as the Bee Line Buzz operation foundered – withdrawals of minibuses started in 1989 with 79 two- and three-year-old buses being taken out of service. Most quickly found new homes elsewhere. The early Dodges which had wide doors and 18 seats (compared with 20 on later vehicles) were among the first to go.

The post-deregulation period saw quick changes in management and in liveries at GM Buses. The new company was led into the deregulation fray by managing director Ralph Roberts, who had joined it in 1985 from NBC. GMT director general David Graham elected to join the new PTE rather than the bus operating company. Roberts retired in October 1988 after three of the toughest years in the history of Manchester bus operation and was succeeded by his recently-appointed deputy, Eric Burling, who replaced the three areas with the new depot-based management structure. Burling was not a bus man and after little more than 12 months resigned, being replaced by Alan Westwell. Westwell had been in charge of Strathclyde Buses, the former PTE bus company in Glasgow, where he had led a spirited and successful defence of the company's operations under concerted attack from the Scottish Bus Group.

In December 1988 the government asked the PTA what plans it had for the future structure and ownership of GM Buses, which provoked a predictable response: that the company should stay as a single operating unit in public ownership. PTA chairman Guy Harkin said: "Any attempt to break up GM Buses would result in substantial additional costs for operators and users. This means higher fares and a poor level of service." At this stage GM Buses had 60 competitors. Harkin argued that breaking the company up would destroy the three years work spent trying to rebuild the bus network since the initial chaos of deregulation. "Any suggestion that GM Buses is too large and has undermined competition is clearly ridiculous", he continued. "The PTA will fight tooth and nail to maintain GM Buses in its present form and in the public sector."

The government's response was to announce that it would still consider splitting the company up, and to examine the feasibility of this it appointed a team of investigators who were to report back to the transport minister by March 1990. No report was ever made public.

The outcome was continuing uncertainty as GM Buses became a political football, with the Labour-led local politicians and Conservative central government playing a confrontational game which appeared to have much more to do with politics than the needs of Greater Manchester's travelling public or GM Buses' employees.

One of GM Buses' three Ailsas, 1446, shows how the original area identities were removed. The stripe between the wheelarches has gone and the name has been removed from above the entrance door.

A statement from the PTA and GM Buses said that the company's share of total bus mileage in the area had fallen from over 95 per cent before deregulation to 87 per cent immediately after deregulation and to 67 per cent by the end of 1988. Commercial competition on Manchester's congested streets was mirrored in competition for PTE tendered services. The number of bids per tender rose from 2.5 in 1987 to 4.0 in the latter part of 1988, and GM Buses was losing out on tenders too, as the average price quoted for tenders fell from 91p a mile in 1986 to 83p a mile in 1989. Noting this trend, GM Buses commented drily in its 1987 annual report: "There is evidence to suggest that some contracts have been won by competitors at prices which cannot realistically provide for the replacement of assets".

Greater Manchester had more competing bus operators than any other metropolitan area in England. The PTA claimed in the spring of 1989 that the figures for operators in each area were:

Greater Manchester	63
South Yorkshire	47
West Midlands	38
West Yorkshire	36
Tyne and Wear	26
Merseyside	25

While it is hard not to view the changes to Manchester's bus operations with some pessimism, things were altogether brighter on the rail front. In the spring of 1987, while the light rapid transit proposals were still awaiting parliamentary approval, the PTE borrowed a two-car train due to be delivered to London's new Docklands Light Railway. This was used for a public exhibition at Debdale Park, designed to give people some idea of what LRT was all about.

There were modest events in 1987-88, like the opening of new railway stations at Hag Fold (Atherton), Lostock (Bolton) and Salford Crescent, and the closure of Royton, plus the commissioning of the new Bolton interchange (and a new Wigan bus station). But these were all cast in the shade by the granting of Royal Assent to the Bill authorising the first stage of the new Metrolink.

After all the dashed hopes of the 1970s, when the Picc-Vic link was on and then off again, Manchester was at last to have some money injected into its railway network. And, as if that were not excitement enough, in 1989 the Government also sanctioned the construction of a new rail link to Manchester's rapidly-expanding airport. 1989 also saw the opening of the Windsor Link (but without the accompanying electrification of the Preston line, as originally proposed at the start of the decade) and the launch of British Rail's Network Northwest.

Centre **The new name was applied to buses in the previous livery, as illustrated by Leopard 438 in St Helens. It has a Plaxton body and was originally an LUT bus.** G R Mills

Right **The single-deck version of the 1988 livery was a simple split at waist level with white above and orange below. Another former LUT Leopard, 454, heads for Eccles.** Stewart J Brown

Above **The new livery used a much brighter shade of orange, with white relief for the top deck window surrounds and roof, and the area around the windscreen. A new GM Buses fleetname was introduced with the subtitle 'People on the move'. This was carried on the front upper-deck window, as well as on the body sides. Metrobus 5130 carries the new livery and is followed by 5015 in the previous style from which the wheelarch relief colour has been removed.** Stewart J Brown

The 1990s: Driving Towards the Private Sector

One area of the GM Buses business which was relatively unaffected by the massive changes taking place was its Charterplan coaching operation. True, three years passed with no additions to the fleet, but starting in 1989, Charterplan began buying a mixture of new and modern used coaches. Four were purchased in 1989 – one new Leyland Tiger (17), Charterplan's last, was accompanied by a trio of second-hand Volvo B10Ms (14-16), one from Excelsior of Bournemouth and two from Park of Hamilton. All four coaches had Plaxton Paramount bodywork. A further pair of ex-Park B10Ms followed in 1990 (18/9) along with a new Setra S215HD (20). Two second-hand Setras (22/3) were purchased in 1991; they had previously been operated by Craiggs of Amble, Northumberland. During 1991 five DAF coaches were hired from Hughes DAF – Plaxton-bodied E328EVH, E647KCX and F261/5RJX plus Van Hool-bodied F658LHD.

Top **The last Leyland coach for Charterplan was Tiger 17. It was delivered in 1989 and had Plaxton Paramount 3500 bodywork finished in the new coach livery.** G R Mills

Above **Second-hand Volvos and Setras were added to the Charterplan fleet between 1989 and 1991. The Setras came from Craiggs of Amble and were S215HRs.** Stewart J Brown

The rebuilt Fleetline single-decker, 1697, operated from Bury until 1990. An emergency window was fitted on the nearside, while at the rear engine shrouds were added to tidy the appearance and add strength to the structure.
Stewart J Brown

Below As minibuses became due for repaint GM Buses abandoned the complex Little Gem scheme and reverted to a simple livery similar to that worn by the first Dodges back in 1986. Dodge 1966 in Oldham in 1993 shows the new livery which featured black window surrounds. A number of Dodges carried illuminated headboards with 'I'm a Little Gem' lettering.
Stewart J Brown

Coach re-registering continued. In 1990 the registration from Setra 31 – 583TD – was transferred to Tiger 64 which had been KGS491Y. The Setra became YNE192Y and was then sold. This was only a prelude to a wholesale re-registering in 1991 as follows:

Fleet no.	Old reg no.	New reg no.
1	B368VBA	OIJ9451
2	B369VBA	OIJ9452
3	B370VBA	NIW1673
5	C167ANA	OIJ9455
6	C706END	NIW1676
7	C707END	IIL1947
9	C309ENA	NIW2399
11	C311ENA	SIB1361
12	C312ENA	SIB1832
14	C106AFX	SIB2014

The new registrations were all Ulster marks and the last digit matched the fleet number. Of the missing coaches, Tiger 4 already carried a dateless number but Tigers 8 and 10 were left with their original C-prefix registrations.

The missing fleet number in the Charterplan series, 21, was an MCW Metrorider which had been rebuilt as a 19-seat coach and transferred from the main bus fleet where it had been 1639. It was returned to the bus fleet at the end of 1992. Further re-registering took place in 1993:

Fleet no.	Old reg no.	New reg no.
8	C308ENA	OIW1608
15	E574UHS	PXI7915
16	E578UHS	PXI8916
17	F853JVR	OIW1317
18	F33HGG	OIW1318
19	F35HGG	OIW1319

Three buses were fitted with coach seats in 1990 – Nationals 177 and 190 as 42-seaters and Dominator 2015 as a 70-seater (43/27) with luggage space in the lower saloon. And two buses were re-engined in 1990. One National, 202 at Bolton, was repowered with an 11.6-litre DAF unit, while Olympian 3301 was fitted with a 12.7-litre Gardner LG1200, a rare power unit in a British bus.

The company's next new buses were eight Renault S75 midibuses with Northern Counties bodies fitted with wheelchair lifts (1721-8); they were delivered in 1990-91, primarily for use on Localine services in Wythenshawe (where they replaced the two Leyland Cubs, 1700/1) and Bury. At the same time an order for new big buses was placed which saw something of the old GMT spirit insofar as the decision was made to order models from different manufacturers for evaluation. Initially the order called for 30 double-deckers of four different types – 10 Dennis Dominator, 10 Volvo Citybus, five Leyland Olympian and five Scania N113. The Olympian order was soon changed to five Volvo B10M single-deckers. All were to be bodied by Northern Counties.

Variety in new bus orders was matched by variety in withdrawals and while there was still some pruning of the fleet which mainly affected standard Atlanteans and Fleetlines, some non-standard types vanished too. The last of the Seddon midis had been taken out of service towards the end of 1987. In 1990 the three Ailsas, 1446-8 at Wigan, were withdrawn, along with the three rare double-deck Dennis Falcons, 1471-3. The unique single-deck Fleetline rebuild, 1697, was also withdrawn. The Ailsas and the Fleetline were sold for further service, while the Falcons were stored. Two of the Falcons, 1471/3, were reinstated for service at Stockport at the end of 1993.

Alan Westwell's arrival saw a reappraisal of the livery and the introduction of a richer shade of orange with black skirt and black side window surrounds on both decks. In fact, although it looked black, the relief was actually a very dark shade of grey. White relief was retained for the windscreen surrounds and the area above the top-deck waist. This attractive layout became the fleet standard, and featured on 1991 deliveries. The opportunity was also taken to simplify the minibus livery, which from 1991 became orange lower panels, white above the waist and black skirt and window surrounds.

Delivery of the 25 new double-deckers (which cost £2.5million) started with the Dominators, 2031-40, which arrived in the spring of 1991. These were followed by the Scanias (1463-7) and the Citybus double-deckers (7001-10), two of which were fitted with wheelchair lifts. The first single-deck Volvo, with Northern Counties' new Countybus body (701), was displayed at Coach & Bus 91 at the National Exhibition Centre, along with Volvo 7009, before entering service at Wigan in 1992. In the event it remained unique; four other single-deckers were cancelled. 701 had moulded plastic seats.

The Dominators, with turbocharged Gardner 6LXCT engines, were allocated to Princess Road. It was decided to concentrate vehicle types at particular depots and to this end earlier Dominators 2006-10 and 2026-30 were transferred in to Princess Road from Bolton; 2001-05 and 2021-25 from Rochdale, and 2011-20 moved in from Altrincham. The new Scanias were allocated to Hyde Road, where they were joined by 1461/2 from Atherton. All of the new vehicles had Voith gearboxes – another Westwell innovation – four-speed D854s in the Dennises and Volvos and three-speed D863s in the Scanias.

The bodies incorporated features to meet the recommendations of the government's Disabled Persons Transport Advisory Committee with bright orange handrails, yellow and black step edgings, palm-operated bell pushes, textured grey floor covering around the platform and yellow-on-black destination displays with a new lettering style. The yellow lettering for destination blinds had started to appear on existing GM Buses vehicles in 1990 and was a requirement for buses being operated on PTE tendered services. They also had two-piece instead of four-piece doors and a new grey and blue moquette.

Although still committed to double-deck operation, a number of single-deck types were examined by GM Buses, including a Scania N113 with Plaxton's new Verde body (H912DRO), a South Yorkshire Transport Volvo B10M/Alexander PS-type (J690XAK), a pair of Leyland Tigers with Alexander (Belfast) Q-type bodies for Shearings (H86DVM) and Ulsterbus (TXI1351) and a B10M with Plaxton Bustler bodywork (H158AKU). None were used in service. Strathclyde Buses' Volvo Citybus A602TNS also made a brief visit to Hyde Road.

Northern Counties was not without troubles of its own at this time. Back in 1986 there had been talk of making Northern Counties part of the GM Buses business. That did not happen. Still owned by the PTA, the rapid fall-off in orders following deregulation had led to it running up losses and ultimately going into administrative receivership in the spring of 1991. This arrangement allowed it to continue trading – and to complete bodies for GM Buses and other operators. The link between bus operator and bus builder was finally cut in August 1992 when Northern Counties was sold to its management. Before this happened GM Buses was considering the products of alternative bodybuilders, and in particular Alexander – but in the end it stayed loyal to its traditional supplier.

Top **Ten new Dennis Dominators were delivered in the spring of 1991. On these the livery was further modified with a richer, less garish orange, black window surrounds and skirt, and a modified fleetname with the word 'buses' in black, rather than orange.** Stewart J Brown

Above **The 10 mid-engined Volvo Citybuses started a new numbering series at 7001, echoing the first standards of almost 20 years earlier. They had Northern Counties bodies. 7001 is seen on the Bury to Manchester rail replacement service during the construction of Metrolink.** Stewart J Brown

Right **Completing the intake of new double-deckers were five Scania N113s, also bodied by Northern Counties. These were allocated to Hyde Road, and were joined by the two BR112s which were running from Atherton.** Stewart J Brown

The contract to build the Metrolink, to run from Bury to Altrincham via Manchester city centre, was won by the GMA Group which comprised GEC, AMEC, Mowlem, GMPTE – and GM Buses. The lines to Bury and to Altrincham were existing British Rail links, and both were closed for around six months at the end of 1991 and the start of 1992 to allow essential engineering work to be carried out. New rails were laid through the city centre to connect the previously separate routes, and a spur line was built from Piccadilly Gardens to Piccadilly station.

The first stage, from Victoria to Bury, was in fact originally intended to open in the autumn of 1991. That, and successive optimistic target dates, passed with no sign of the Italian-built trams on the city's streets. Delivery delays and the need for extra staff training to cope with emergencies saw the opening of the Bury line deferred to 23rd March 1992. Only days before the big date it was delayed by another two weeks to 6th April. The service initially ran from Victoria; on-street running through the city centre to the G-Mex exhibition centre started on 27th April. The Altrincham section followed on 15th June, and the final section, the spur to Piccadilly station, was opened on 20th July. The £130 million system was formally opened by the Queen on 17th July and was soon carrying 25,000 people daily.

The 29m long articulated cars were built in Italy by the Firema Consortium and weighed 45 tonnes unladen. Original PTE publicity included artist's impressions showing a white and orange livery – but the reality was a bland grey and blue. Each articulated unit had 86 seats and capacity for a theoretical 188 standees, although Metrolink was claiming a more modest 120. There were four doors on each side and driving cabs at each end. The maximum running speed was 80km/hr on reserved track which was reduced to 48km/hr for on-street running. There were two power bogies, with a third bogie placed centrally under the articulation.

Although operated and maintained by Greater Manchester Metro Ltd until 2007, ownership of the system remains with the PTE.

As soon as Metrolink was operational, the PTA was announcing plans to extend it. It had approval for lines to Salford Quays and Trafford Park – provided it could raise the necessary money – and it also wanted lines to Ashton-under-Lyne and to Wythenshawe and the Airport. The last-named was rendered unnecessary by a new British Rail link to the Airport which opened in the spring of 1993 – and also killed off GM Buses' Airport Express service.

Ironically GM Buses, which had been providing a rail-replacement service between Bury and Manchester with its new Volvo Citybuses while the line was closed, announced that it would retain the route in competition with Metrolink, offering lower fares (£9.99 for a weekly ticket against £16.20 on Metrolink) but slower journeys – 40 minutes against 25. When the Metrolink service started the minimum fare, even for a short hop across central Manchester, was a high 70p.

Metrolink started running in 1992, with Italian-built articulated light rail vehicles linking Bury with Altrincham. The first unit, 1001, makes its way through central Manchester. Stewart J Brown

Latterly the airport service was branded as the Airport Shuttle and the route number changed from 200 to 757. The livery was also modified. The service was withdrawn in 1993 after the opening of the rail link to the airport. Stewart J Brown

While Metrolink had an assured future and adequate funding despite the Conservative government's apparent apathy towards public transport, no such security was afforded to GM Buses. The government was anxious to see the former PTE-owned bus companies being privatised and between 1988 and 1992 three were – Yorkshire Rider in West Yorkshire, Busways in Tyne & Wear and West Midlands Travel. The other three English operators affected – GM Buses, Merseybus and South Yorkshire Transport – were controlled by PTAs which were less willing to sell their bus operations. The politicians in Greater Manchester were openly opposed to any sell-off.

Whatever wrangles there may have been about the company's ownership, it was still suffering at the hands of a myriad of competitors and in August 1991 announced a new round of cuts – mainly in overheads rather than in service levels. This involved the closure of four depots in November – Rochdale, Swinton, Tameside (opened in 1977) and Altrincham (opened in 1981). GM Buses was left with nine depots, at Queens Road, Princess Road, Hyde Road, Oldham, Atherton, Wigan, Bury, Bolton and Stockport – and the outstation at Glossop. There had been 19 only six years earlier. Plans were also drawn up to close the company's head offices in Devonshire Street North (inherited from Manchester City Transport) and to run the company with a streamlined management structure from the Charles Street offices in Stockport. This did not happen.

Low-cost operating units with lower pay rates were set up in the spring of 1992 running seven Metroriders and seven Atlanteans from Bennett Street (site of the company's training school, behind Hyde Road depot, and given depot code BT) and 13 Dodges from the Charterplan garage at Charles Street (code CS) in Stockport. At the same time a fleet of 10 MCW Metroriders – later increased to 17 – was hired from West Midlands Travel and operated on services to compete with Citibus – running in full WMT silver and blue livery. This brought complaints about GM Buses' competitive tactics. Citibus, whose buses were blue, argued that the WMT buses would confuse Citibus passengers. The Metroriders were quickly given GM Buses fleetnames and some were ultimately repainted in fleet livery. Additional low-cost special operating units were set up in 1992-93 at Atherton (SA, South Atherton), Oldham (MS, Mumps), Boyle Street, Manchester (BE), and Bolton (SB, South Bolton), as GM Buses tried to fend off continuing high levels of competition.

Two very different demonstrators examined in the autumn of 1992 were J366BNW, an Optare MetroRider, and SUG561M, an East Lancs Atlantean Sprint. The MetroRider, which was an update of the defunct MCW model already in use with GM Buses, was tried in service. The Sprint, basically nothing more than an 18-year-old Atlantean with a new East Lancs single-deck body, wasn't. Neither type was bought by GM Buses.

Withdrawals in 1992 saw the last of the Iveco Ford minibuses go, and the operational Leyland National fleet being cut to just 13 buses. At the same time the biggest-ever fleet renumbering was implemented, with all buses in the 6xxx and 8xxx series being renumbered in a new 4xxx series. Thus all surviving standards between 8002 and 8765 became 4002 to 4765, while former LUT Fleetlines in the range 6942 to 6990 became 4942 to 4990. The change brought fleet numbers into line with a new radio system and its call signs, but it took many months before all affected buses had their new numbers applied.

To speed the introduction of the revised livery many buses received partial repaints, usually to apply the dark grey skirt. A minor modification introduced in 1993 was the deletion of the white windscreen and headlamp surrounds.

1993 was to be the turning point for GM Buses. There had been a general election in 1992 which saw the Conservatives re-elected – and which dashed any hopes harboured by Labour politicians in Manchester and elsewhere that they might be able to preserve public ownership of their bus fleets. In November 1992 transport secretary John McGregor announced that he intended to force the split-up of GM Buses, using powers which he had under the 1985 Transport Act. And, in a statement which bore little resemblance to reality, he said: "The very size of GM Buses has tended to inhibit competition between operators and I am convinced that dividing the company into smaller units will bring better services for passengers."

The PTA responded angrily, saying that it had previously indicated to the government its willingness to sell the company to its employees. Councillor Joe Clarke, chairman of the PTA, demanded the resignation of public transport minister Roger Freeman, saying that he had gone back on commitments to hold further discussions. The government did not indicate how many parts it wanted GM Buses to be divided into – the favoured options were thought to be either two or three. "We already have over 68 bus operators in the county – why does Greater Manchester want 71 operators?" Clarke argued, saying of the government: "Perhaps it wants to sell the parts to the company's competitors who regularly beat a path to the minister's door."

Most of GM Buses' 4,800 employees signed a petition opposing the break-up and this was delivered to the prime minister, John Major, in January 1993. But to no avail. By April it was becoming clear that the company would be split, with the PTA reluctantly proposing the formation of two new operat-

The last vehicles bought by GM Buses before it was split were seven London Buses MCW Metroriders. 1682, formerly London MR48, is seen outside Manchester's Arndale bus station. Stewart J Brown

Metroriders were hired from West Midlands Travel in 1992 and were operated in WMT colours. After protests from competitor Citibus, the Metroriders had GM Buses fleetnames added. Stewart J Brown

Left **More new names soon replaced plain GM Buses.** Stewart J Brown

The sale – due to be completed by December 1993 – was further delayed. The two companies officially became operational, still in public sector ownership, on 13th December.

There was trouble at GM Buses (South) too. Stagecoach Holdings, one of the bidders for the company, bid again – and even advertised in the Manchester Evening News in an effort to get the GM Buses (South) employees to switch their support from a management-led buy-out to a Stagecoach purchase. Stagecoach said if it bought the company it would invest in 125 new single-deck buses within 12 months. However Stagecoach's overtures were rejected and it started running competing services under the Stagecoach Manchester name with a fleet of brand new Alexander Dash midibuses. This operation was controlled by Ribble.

While this was going on, the PTA agreed to accept a management/employee buy-out offer for GM Buses (North), much to the chagrin of British Bus. West Midlands Travel – the privatised successor to the West Midlands PTE bus operation – indicated that it wanted to bid for the North company too, but in the end the sale to the employees went ahead. While still owned by the PTA GM Buses (North) bought five second-hand buses from Yorkshire Rider – one Fleetline and four Atlanteans which were making their way back across the Pennines. The Fleetline, PTD642S, had originally been LUT 6914 and was numbered 4914. The four R- and S-registered Atlanteans were one-time GMT vehicles and received their original fleet numbers once again – 7685, 7752/62/73.

And so both GM Buses (South) and GM Buses (North) became employee-owned on 31st March 1994. GM Buses (South), headed by chairman Ken Harvey, nominated by the company's fund-holders, and managing director Peter Short, took over 795 buses operated by 2,000 employees. Its depots were Hyde Road and Princess Road in Manchester plus Stockport and Glossop. Included in the sale were Charterplan and GM Engineering, formed to run as a central engineering function able to provide services to outside companies. GM Buses (North) had Gavin Laird, former leader of the boilermakers union, as its chairman with Alan Westwell as managing director. It had 2,500 employees and 965 buses running from depots in Atherton, Bolton, Bury, Oldham, Wigan and Queens Road, Manchester.

The conditions facing GM Buses and its two successors in the 1990s were a far cry from the high ideals of the late 1960s. It is hard to believe that in a quarter of a century which has seen traffic congestion and vehicle pollution worsen, an efficient and co-ordinated public transport system could have been systematically undermined in the pursuit of political dogma. GM Buses had an unenviable task in maintaining service quality and staff morale in an era when low-cost operators were attacking it from all quarters. Selnec and GMT may not have been perfect – but from the viewpoint of the 1990s, theirs were the halcyon days of bus operation in Greater Manchester.

ing units. The north unit would take over Atherton, Bolton, Bury, Oldham, Queens Road and Wigan. The south unit would run Glossop, Hyde Road, Princess Road and Stockport.

The north unit was put under the stewardship of managing director Alan Westwell and operations director Rodney Dickinson. The south unit was in the hands of finance director Peter Short and commercial director Les Wheatley. Both groups set to work evolving plans for management/employee buy-outs for what became known as GM Buses (North) and GM Buses (South).

While all this was going on, the competitive battle had still to be fought. In the summer of 1993 British Bus and Merseyside Transport struck a deal which was to see Merseyside take over the Bee Line operations in Rochdale. That foundered at the last minute and instead Merseyside set up its own operation in the area, MTL Manchester, serving Rochdale, Bury, Oldham and Manchester. GM Buses reacted by launching a new service in Liverpool, running to Croxteth and competing head on with Merseyside. The sight of two erstwhile PTE operators engaged in tit-for-tat bus wars seemed to be the final ignominy in the brave new world of free market public transport.

The last buses to join the GM Buses fleet did so in late summer 1993. They were not smart new £100,000 double-deckers. Nor were they marginally less-expensive new single-deckers. They were seven MCW Metroriders, 1681-87, all six years old and bought second-hand from London Buses – the capital's third contribution to GM Buses and its predecessors following the 20 LUT DMSs and the 10 short-lived Routemasters.

It was not until November 1993 that an announcement was made about GM Buses' fate. A number of organisations had expressed interest in taking over the two businesses and in the end the PTA announced that GM Buses (South) was to go to a management-led employee buy-out team, while GM Buses (North) was being sold to British Bus, which already had a significant presence in the conurbation.

This decision brought a storm of protest from supporters of a management/employee buy-out for GM Buses (North). They argued, quite reasonably, that if one of the aims of the split-up and sale of GM Buses was to end large fleet domination there was little logic in selling to British Bus. British Bus owned North Western in Liverpool, Bee Line in Manchester and Midland Red North which served Cheshire.

FLEET LIST of SELNEC and GREATER MANCHESTER PTE BUSES

Central Area single-deckers

1-5	9746-9750NA	Leyland Tiger Cub PSUC1/2	Park Royal	DP40F	1961
6-9	3651-3654NE	Leyland Tiger Cub PSUC1/12	Park Royal	DP40F	1962
10-15	3655-3660NE	Leyland Tiger Cub PSUC1/12	Park Royal	DP38D	1962
16-17	ANF161-162B	Leyland Panther Cub PSRC1/1	Park Royal	B43D	1964
18-35	BND863-880C	Leyland Panther Cub PSRC1/1	Park Royal	B43D	1965
36-54	GND81-99E	Leyland Panther PSUR1/1	Metro-Cammell	B44D	1967
55-64	GND101-110E	Leyland Panther PSUR1/1	Metro-Cammell	B44D	1967
65-73	TRJ102-110	AEC Reliance 2MU3RV	Weymann	B45F	1962

1-64 were Manchester 46-99, 101-110.
65-73 were Salford 102-110.
10-15 were renumbered 5000-5 and transferred to Selnec Southern.

Mancunians

1001-1048	HVM901-948F	Leyland Atlantean PDR1/1	Park Royal	H45/28D	1968
1051-1097	LNA151-197G	Leyland Atlantean PDR2/1	Park Royal	H47/29D	1968-69
1101-1126	NNB510-535H	Leyland Atlantean PDR2/1	Park Royal	H47/28D	1970
1131-1142	NNB536-547H	Leyland Atlantean PDR2/1	East Lancs	H47/32F	1969
1143-1154	NNB548-559H	Leyland Atlantean PDR2/1	East Lancs	H47/26D	1969
1161-1194	ONF849-882H	Leyland Atlantean PDR2/1	Park Royal	H47/28D	1970
1201-1220	SRJ324-343H	Leyland Atlantean PDR2/1	Metro-Cammell	H47/31D	1970

1001-48, 1051-97 were ex Manchester (same numbers). 1044 was out of service after an accident and did not run for Selnec.
The 11xx series buses were ordered by Manchester. 1201-20 were ordered by Salford, and would have been numbered 324-343 in the Salford fleet. Buses from 1161 were delivered in Selnec livery.

2001-2048	HVM801-848F	Daimler Fleetline CRG6LX	Park Royal	H45/28D	1968
2051-2092	LNA251-292G	Daimler Fleetline CRG6LX	Park Royal	H47/28D	1969
2093-2097	LNA293-297G	Daimler Fleetline CRG6LX	Park Royal	H47/28D	1969
2101-2144	NNB560-603H	Daimler Fleetline CRG6LXB	Park Royal	H47/28D	1970
2151-2161	ONF883-893H	Daimler Fleetline CRG6LXB	Metro-Cammell	H47/30D	1970
2162-2210	PNA201-249J	Daimler Fleetline CRG6LXB	Metro-Cammell	H47/30D	1970-71
2211-2270	RNA211-270J	Daimler Fleetline CRG6LXB	Park Royal	H47/29D	1971
2271-2304	SVR271-304K	Daimler Fleetline CRG6LXB	Roe	H43/32F	1972

2001-48, 2051-97 were ex Manchester (same numbers); buses from 2101 were ordered by Manchester. 2101-44 were delivered in Manchester livery; buses from 2151 were delivered in Selnec livery.
2271-2304 were originally ordered with East Lancs bodies.

3000 series: Selnec Central Leyland double-deckers

3000-3001	TRJ149-150	Leyland Atlantean PDR1/1	Metro-Cammell	H44/33F	1962
3002-3039	WRJ151-188	Leyland Titan PD2/40	Metro-Cammell	H36/28F	1963
3040-3054	ARJ191-205B	Leyland Titan PD2/40	Metro-Cammell	H36/28F	1964
3055-3057	ARJ209-211B	Leyland Atlantean PDR1/1	Metro-Cammell	H43/33F	1964
3058-3078	DBA212-232C	Leyland Atlantean PDR1/1	Metro-Cammell	H43/33F	1965
3079-3103	FRJ233-257D	Leyland Titan PD2/40	Metro-Cammell	H36/28F	1966
3104-3128	JRJ258-282E	Leyland Titan PD2/40	Metro-Cammell	H36/28F	1967
3129-3149	MRJ283-303F	Leyland Atlantean PDR1/1	Metro-Cammell	H44/33F	1968
3150-3169	PRJ304-323G	Leyland Atlantean PDR1A/1	Park Royal	H43/29D	1969

3000-3169 were Salford 149-188, 191-205, 209-323.

3200-3223	JND601-624	Leyland Titan PD2/3	Metro-Cammell	H32/26R	1951
3225-3264	JND626-665	Leyland Titan PD2/3	Metro-Cammell	H32/26R	1951
3287	JND688	Leyland Titan PD2/3	Leyland	H32/26R	1950
3294	JND695	Leyland Titan PD2/3	Leyland	H32/26R	1950
3299	JND700	Leyland Titan PD2/3	Leyland	H32/26R	1950
3323	NNB163	Leyland Titan PD2/12	Northern Counties	H33/28R	1954
3331-3332	NNB171-172	Leyland Titan PD2/12	Leyland	H32/28R	1953
3334	NNB174	Leyland Titan PD2/12	Leyland	H32/28R	1953
3337	NNB177	Leyland Titan PD2/12	Leyland	H32/28R	1953
3339-3340	NNB179-180	Leyland Titan PD2/12	Leyland	H32/28R	1953
3342	NNB182	Leyland Titan PD2/12	Leyland	H32/28R	1953
3345-3347	NNB185-187	Leyland Titan PD2/12	Leyland	H32/28R	1953
3350-3352	NNB190-192	Leyland Titan PD2/12	Leyland	H32/28R	1953
3354	NNB194	Leyland Titan PD2/12	Leyland	H32/28R	1953
3356-3360	NNB196-200	Leyland Titan PD2/12	Leyland	H32/28R	1953
3364	NNB204	Leyland Titan PD2/12	Leyland	H32/28R	1953
3411-3470	PND411-470	Leyland Titan PD2/12	Metro-Cammell	H36/28R	1956
3471-3493	TNA471-493	Leyland Titan PD2/40	Burlingham	H37/28R	1958
3494	TNA494	Leyland Titan PD2/40	Leyland (a)	H32/28R	1958
3495-3514	TNA495-514	Leyland Titan PD2/40	Burlingham	H37/28R	1958
3515-3520	TNA515-520	Leyland Titan PD2/34	Burlingham	H37/28R	1958
3521-3620	UNB521-620	Leyland Titan PD2/40	Metro-Cammell	H37/28R	1958-59
3621-3630	UNB621-630	Leyland Atlantean PDR1/1	Metro-Cammell	H43/34F*	1960
3631-3670	9831-9870NA	Leyland Titan PD2/37	Metro-Cammell	H37/28R	1961
3671-3695	3671-3695NE	Leyland Titan PD2/37	Metro-Cammell	H37/28R	1963
3696	889VU	Leyland Titan PD2/37	Metro-Cammell	H37/28R	1964
3697-3720	3697-3720VM	Leyland Titan PD2/37	Metro-Cammell	H37/28R	1964
3721-3772	BND721-772C	Leyland Atlantean PDR1/2	Metro-Cammell	H43/33F*	1965
3773-3792	BND773-792C	Leyland Atlantean PDR1/2	Metro-Cammell	H43/32F	1965
3801-3860	END801-860D	Leyland Atlantean PDR1/2	Metro-Cammell	H43/32F	1966

Buses in the series 3200-3860 were ex Manchester where they had the same fleet numbers.
(a) 1953 Leyland body transferred from 3363.
*3629 was H43/33F; 3721-72 were later reseated to H43/32F.

4000 series: Selnec Central Daimler double-deckers

4000-4029	TRJ111-140	Daimler CVG6	Metro-Cammell	H37/28R	1962
4030-4035	TRJ141-146	Daimler CVG6	Metro-Cammell	H36/28F	1962
4036-4037	TRJ147-148	Daimler Fleetline CRG6LX	Metro-Cammell	H44/31F	1963
4038-4039	ARJ189-190B	Daimler CCG6	Metro-Cammell	H36/28F	1964
4040-4042	ARJ206-208B	Daimler Fleetline CRG6LX	Metro-Cammell	H44/33F	1964
4043-4044	CRJ415-416	Daimler CVG6	Metro-Cammell	H30/24R	1950
4045-4046	CRJ418/25	Daimler CVG6	Metro-Cammell	H30/24R	1950
4047-4048	CRJ428-429	Daimler CVG6	Metro-Cammell	H30/24R	1950
4049-4050	CRJ433/9	Daimler CVG6	Metro-Cammell	H30/24R	1951
4051	FRJ457	Daimler CVG6	Metro-Cammell	H30/24R	1951
4052-4053	FRJ461/3	Daimler CVG6	Metro-Cammell	H30/24R	1951
4054-4055	FRJ465/70	Daimler CVG6	Metro-Cammell	H30/24R	1951
4056-4057	FRJ473/8	Daimler CVG6	Metro-Cammell	H30/24R	1951
4058-4060	FRJ483-485	Daimler CVG6	Metro-Cammell	H30/24R	1951
4061-4062	FRJ488/98	Daimler CVG6	Metro-Cammell	H30/24R,	1951
4063-4064	FRJ506-507	Daimler CVG6	Metro-Cammell	H30/24R	1951
4065	FRJ511	Daimler CVG6	Metro-Cammell	H30/24R	1951
4066-4067	FRJ521-522	Daimler CVG6	Metro-Cammell	H30/24R	1951
4068-4069	FRJ524-525	Daimler CVG6	Metro-Cammell	H30/24R	1951
4070-4072	FRJ527-529	Daimler CVG6	Metro-Cammell	H30/24R	1951
4073-4074	FRJ531/3	Daimler CVG6	Metro-Cammell	H30/24R	1951
4075-4081	FRJ535-541	Daimler CVG6	Metro-Cammell	H30/24R	1951
4082-4084	FRJ543-545	Daimler CVG6	Metro-Cammell	H30/24R	1951
4085-4086	FRJ547-548	Daimler CVG6	Metro-Cammell	H30/24R	1951
4087-4089	FRJ552-554	Daimler CVG6	Metro-Cammell	H30/24R	1951
4090	FRJ560	Daimler CVG6	Metro-Cammell	H30/24R	1951

4000-4090 were ex-Salford. Their Salford fleet numbers were the same as the registration numbers.

4111/8	JND712/9	Daimler CVG6K	Metro-Cammell	H32/26R	1950
4122-4137	JND723-738	Daimler CVG6K	Metro-Cammell	H32/26R	1950
4139-4148	JND740-749	Daimler CVG6K	Metro-Cammell	H32/26R	1951
4150-4174	KND911-935	Daimler CVG6K	Metro-Cammell	H32/26R	1951
4176-4189	KND937-950	Daimler CVG6K	Metro-Cammell	H32/28R	1951
4400-4479	NNB210-289	Daimler CVG6K	Metro-Cammell	H33/28R	1953-54
4480-4489	NNB290-299	Daimler CLG5K	Metro-Cammell	H36/28R	1955
4490	PND490	Daimler CVG5K	Metro-Cammell	H36/28R	1955
4491-4509	PND491-509	Daimler CVG5K	Metro-Cammell	H36/28R	1956-57
4510-4549	RND510-549	Daimler CVG6K	Northern Counties	H37/28R	1957-58
4550-4579	TNA550-579	Daimler CVG6K	Burlingham	H37/28R	1961
4580-4589	9580-9589NA	Daimler CVG6K	Metro-Cammell	H37/28R	1962
4590-4609	4590-4609NE	Daimler Fleetline CRG6LX	Metro-Cammell	H43/33F	1963
4610-4629	4610-4629VM	Daimler Fleetline CRG6LX	Metro-Cammell	H43/33F	1963
4630-4649	4630-4649VM	Daimler CVG6K	Metro-Cammell	H37/29R	1963
4650-4654	4650-4654VM	Daimler CCG6K	Metro-Cammell	H37/28R	1964
4655-4684	ANA655-684B	Daimler Fleetline CRG6LX	Metro-Cammell	H43/33F	1964
4701-4730	DNF701-730C	Daimler Fleetline CRG6LX	Metro-Cammell	H43/32F	1965
4731-4760	FNE731-760D	Daimler Fleetline CRG6LX	Metro-Cammell	H43/32F	1966-67

Buses in the series 4111-4760 were ex Manchester where they had the same fleet numbers.

5000 series: Selnec Southern single-deckers

5000-5005	3655-3660NE	Leyland Tiger Cub PSUC1/12	Park Royal	DP38D	1962
5011-5014	111-114JBU	Leyland Tiger Cub PSUC1/13	Marshall	B41D	1964
5015-5016	115-116JBU	Leyland Tiger Cub PSUC1/13	Pennine	B41D	1965
5017-5020	LBU117-120E	Leyland Panther Cub PSRC1/1	Marshall	B45D	1967
5022-5027	OBU172-177F	Leyland Panther PSUR1/1R	Marshall	B48D	1968
5055-5056	CTC355-356E	Leyland Panther Cub PSRC1/1	East Lancs	DP43F	1967
5068	XLG477	Atkinson PL746H	Northern Counties	B34C	1956
5070-5072	993-995GMA	Atkinson PL745H	Northern Counties	B41F	1959
5073-5075	WMA113-115E	Bristol RESL6G	Northern Counties	B43F	1967
5076-5078	YLG716-718F	Bristol RESL6G	Northern Counties	B43F	1967
5080-5083	NDB353-356	Leyland Tiger Cub PSUC1/1	Crossley	B44F	1958
5084-5088	KDB404-408F	Leyland Leopard PSU4/1R	East Lancs	B43D	1968

5000-5005 were transferred from Central division in November 1970 and were renumbered from 10-15 in April 1973. They were originally Manchester 55-60.
5011-20/2-7 were Oldham 111-120, 172-177.
5055/56 were Ashton 55/6.
5068/70-78 were SHMD 108, 110-118.
5080-5088 were Stockport 400-408.

141

5100-5378: Selnec Southern ex-Oldham double-deckers

5101-5107	101-107HBU	Leyland Titan PD3/5	Roe	H41/32F	1964
5109-5110	109-110HBU	Leyland Titan PD3/5	Roe	H41/32F	1964
5121-5130	CBU121-130C	Leyland Atlantean PDR1/1	Roe	H43/34F	1965
5131-5135	GBU131-135D	Leyland Atlantean PDR1/1	East Lancs	H43/34F	1966
5136-5147	GBU136-147D	Leyland Atlantean PDR1/1	Roe	H43/34F	1966
5148-5152	LBU148-152E	Leyland Atlantean PDR1/1	Neepsend	H43/34F	1967
5153-5160	LBU153-160E	Leyland Atlantean PDR1/1	Roe	H43/34F	1967
5161-5171	OBU161-171F	Leyland Atlantean PDR1/1	Roe	H43/34F	1967-68
5178-5182	SBU178-182G	Leyland Atlantean PDR1A/1	Roe	H43/31D	1969
5183-5187	WBU183-187H	Leyland Atlantean PDR1A/1	Roe	H43/28D	1970
5188-5199	ABU188-199J	Leyland Atlantean PDR1A/1	Roe	H43/31D	1971
5202	EDB556	Leyland Titan PD2/1	Leyland	H30/26R	1951
5241-5242	EBU871-872	Leyland Titan PD2/3	Roe	H31/25R	1950
5245	EBU875	Leyland Titan PD2/3	Roe	H31/25R	1950
5246	DBU246	Leyland Titan PD1/3	Roe	H31/25R	1947
5249	EBU879	Leyland Titan PD2/3	Roe	H31/25R	1950
5260	FBU647	Leyland Titan PD2/3	Roe	H31/25R	1950
5270	HBU123	Leyland Titan PD2/12	Leyland	H30/26R	1952
5273-5277	KBU383-387	Leyland Titan PD2/20	Metro-Cammell	H30/28R	1955
5278-5287	KBU373-382	Leyland Titan PD2/20	Roe	H31/28R	1954
5288-5307	NBU488-507	Leyland Titan PD2/20	Roe	H31/29R	1957
5309	NBU509	Leyland Titan PD2/20	Crossley	H33/28R	1957
5313-5318	NBU513-518	Leyland Titan PD2/20	Northern Counties	H33/28R	1957
5319-5328	PBU919-928	Leyland Titan PD2/30	Metro-Cammell	H37/28R	1958
5329-5352	PBU929-952	Leyland Titan PD2/30	Roe	H37/28R	1958
5353-5362	PBU953-962	Leyland Titan PD2/30	Northern Counties	H37/28R	1958-59
5364-5365	LWE104	Leyland Titan PD2/1	Leyland	H30/26R	1949
	LWE109-110	Leyland Titan PD2/1	Leyland	H33/26R	1949
5371-5372	LWE111	Leyland Titan PD2/1	Leyland	H32/26R	1949
5373-5374	DBN329-330	Leyland Titan PD2/4	Leyland	H32/26R	1949
5375-5376	DBN337/41	Leyland Titan PD2/4	Leyland	H33/28R	1952
5377-5378	OWB856-857	Leyland Titan PD2/10	Leyland	H33/28R	1952
	OWB859/61	Leyland Titan PD2/10	Leyland		

5101-7/9/10, 5121-71/8-82 were Oldham 101-7/9/10, 121-171/8-82.
5183-5187 were ordered by Oldham and were delivered to Selnec in Oldham livery.
5188-5199 were ordered by Oldham and were delivered to Selnec in Selnec livery.
5202 was Stockport 302 and carried this number in error. It was renumbered 5922.
5241/2/5/6/9, 5260, 5270/3-5307/9/13-62 were Oldham 341/2/5, 246, 349/60/70/3-407/9/13-62.
LWE104, 5364/5, LWE111 were Oldham 463-466 and were new to Sheffield, moving to Oldham in 1965.
5371-5374 were Oldham 471-474 and were new to Bolton, joining the Oldham fleet in 1965.
5375-5378 were Oldham 475-478 and were new to Sheffield, joining the Oldham fleet in 1966.

54xx series: Selnec Southern ex-Ashton double-deckers

5407-5410	LTC767-770	Leyland Titan PD3/5	Leyland	H30/26R	1950
5411-5417	UTB311-317	Leyland Titan PD2/12	Crossley	H32/28R	1955
5418-5423	18-23NTD	Leyland Titan PD2/40	Roe	H37/28R	1960
5424-5431	224-231YTB	Leyland Titan PD2/40	Roe	H37/28R	1962
5432-5433	332/8TF	Leyland Titan PD2/40	Roe	H37/28R	1963
5434-5437	334-337TF	Leyland Titan PD2/40	Roe	H37/28R	1964
5438-5441	DTJ138-141B	Leyland Titan PD2/37	Roe	H37/28F	1965
5442-5446	PTE942-946C	Leyland Titan PD2/37	Roe	H43/32F	1966
5447-5454	YTE847-854D	Leyland Atlantean PDR1/1	Northern Counties	H43/28D	1969
5457-5461	PTF857-861G	Leyland Atlantean PDR1/1	Northern Counties	H43/28D	1970
5462-5465	VTE162-165H	Leyland Atlantean PDR1A/1	Bond	H32/28R	1956
5466-5468	XTC853-855	Guy Arab IV 6LW	Bond	H32/28R	1956
5469	XTC852	Guy Arab IV 6LW			
5466-5471	PNF941-946J	Leyland Atlantean PDR1A/1	Northern Counties	H43/32F	1971

5407-54/7-61 were Ashton 7-54, 57-61.
5462-5465 were ordered by Ashton and delivered to Selnec in Selnec livery.
5466-5469 (Guys) were Ashton 66-68, 65.
5466-5471(Atlanteans) had chassis ordered by Ashton. They were previously EX1-6.

56xx series: Selnec Southern ex-SHMD double-deckers

5601-5606	101-106UTU	Leyland Titan PD2/37	Northern Counties	H36/28F	1962
5607-5612	ATU407-412B	Daimler CVG6	Northern Counties	H36/28F	1964
5613-5621	GTU113-121C	Daimler Fleetline CRG6LX	Northern Counties	H43/31F	1965
5622-5637	NMA322-337D	Daimler Fleetline CRG6LX	Northern Counties	H43/31F	1966
5638-5647	ELG38-47F	Daimler Fleetline CRG6LW	Northern Counties	H41/27D	1968
5661-5664	OMB161-164	Daimler CVD6	Northern Counties	H36/28R	1952
	OMB165	Daimler CVD6	Northern Counties	H36/28R	1952
5666	OMB166	Atkinson PD746	Northern Counties	H36/28R	1952
	UMA370	Daimler CVD6	Northern Counties	H35/24C	1955
5671-5676	VTU71-76	Daimler CVG6	Northern Counties	H35/23C	1956
5679-5684	279-284ATU	Daimler CVG6	Northern Counties	H36/28R	1957
5685-5692	85-92ETU	Leyland Titan PD2/40	Northern Counties	H36/28R	1958
5696-5697	696-697GTU	Daimler CSG6	Northern Counties	H36/28R	1959
5698-5699	698-699GTU	Leyland Titan PD2/40	Northern Counties	H36/28R	1959
	EDB551-552	Leyland Titan PD2/1	Leyland	H30/26R	1951

5601-5647 were SHMD1-47.
5661-64, OMB165, 5666, UMA370, 5671-6/9-92/6-9 were SHMD 61-66, 70-76, 79-92, 96-99.
EDB551/52 were Stockport 297/8 and were transferred to Stalybridge by Selnec and allocated SHMD numbers 51/2. 52 was repainted in SHMD livery.

5801-5995: Selnec Southern ex-Stockport double-deckers

5801-5810	YDB1-10	Leyland Titan PD2A/30	East Lancs	H36/28R	1963
5811-5825	BJA911-925B	Leyland Titan PD2/40	East Lancs	H36/28R	1964
5826-5840	FDB326-340C	Leyland Titan PD2/40	East Lancs	H36/28R	1965
5841-5855	HJA941-955E	Leyland Titan PD2/40	East Lancs	H36/28R	1967
5856-5870	HJA956-970E	Leyland Titan PD2/40	Neepsend	H36/28R	1967
5871-5885	KJA871-885F	Leyland Titan PD3/14	East Lancs	H38/32R	1968
5886-5891	MJA886-891G	Leyland Titan PD3/14	East Lancs	H38/32R	1969
5892-5897	MJA892-897G	Leyland Titan PD3/14	East Lancs	H38/32F	1969
5898-5907	allocated to Bristol VRTs destroyed in a fire at East Lancs in April 1970				
5913	EDB547	Leyland Titan PD2/1	Leyland	H30/26R	1951
5917-5927	EDB551-561	Leyland Titan PD2/1	Leyland	H30/26R	1951
5933-5936	NDB366-369	Leyland Titan PD2/30	Crossley	H33/28R	1958
5937-5942	NDB360-365	Leyland Titan PD2/30	Crossley	H33/28R	1958
5943-5952	PJA913-922	Leyland Titan PD2/30	Longwell Green	H32/28R	1960
5953-5962	VDB584-593	Leyland Titan PD2A/30	East Lancs	H32/28R	1962
5995	EDB549	Leyland Titan PD2/1	Leyland	O22/26R	1951

5801-5897 were Stockport 1-97.
5898-5907 had been ordered by Stockport.
5913/7-27/33-62 were Stockport 293/7-307/33-62.
5917/18 were transferred to Stalybridge in March 1970 and allocated fleet numbers 51/2 in the SHMD series. 52 was repainted in SHMD livery.
5922 was transferred to Oldham in February 1970 and repainted in Oldham livery. It ran briefly as 5202 in the ex-Oldham number series.
5995 was Stockport 295. It had been a tree-lopper and joined the Selnec psv fleet in 1972. It was briefly numbered 2995.

6000 series: Selnec Northern single-deckers

6001-6002	187-188LTB	Leyland Tiger Cub PSUC1/2	East Lancs	DP43F	1960
6009	GWH516	Leyland Royal Tiger PSU1/14	East Lancs	B43F	1955
6010	JBN141	Leyland Royal Tiger PSU1/14	Bond	B44F	1956
6012	UWH322	Leyland Leopard L2	East Lancs	DP41F	1962
6016-6020	2116-2120DK	AEC Reliance 2MU2RA	Weymann	B42D	1961
6021	6321DK	AEC Reliance 2MU2RA	East Lancs	B43D	1964
6022-6023	ADK722-723B	AEC Reliance 2MU2RA	East Lancs	B43D	1964
6024-6029	GDK324-329D	AEC Reliance 6MU2RA	Willowbrook	B45F	1966
6030-6033	LDK830-833G	Daimler Fleetline SRG6LW	Willowbrook	B45F	1968
6034-6037	MDK734-737G	AEC Swift MP2R	Pennine	B47F	1969
6038	KKV700G	Daimler Fleetline SRG6LX	Willowbrook	B43D	1970
6040-6049	TDK540-549K	AEC Swift 2MP2R	Pennine	B42D	1971-2
6054-6055	YBN14-15	Leyland Leopard PSU3/4R	East Lancs	B49D	1964
6056-6057	YBN16-17	Leyland Leopard L2	East Lancs	B43D	1964
6059	XBN976L	Seddon Pennine IV: 236	Seddon	DP25F	1972
6060-6064	HTJ131-135F	Leyland Leopard PSU4/2R	East Lancs	B45F	1968
6081	KEN381G	Bedford J2SZ10	Duple Midland	B21F	1969
6082	RJX258	Albion Nimbus NS3AN	Weymann	B31F	1963
6087	TEN887	AEC Reliance 2MU3RV	Alexander Y	B43F	1964
6088	TEN988	AEC Reliance 2MU3RA	Alexander Y	B43F	1964
6089-6091	FEN89-91E	Daimler Fleetline SRG6LX	East Lancs	B41D	1967
6092-6097	KEN292-297G	Daimler Fleetline SRG6LX	East Lancs	B41D	1969

6001-6002 were Leigh 1,2.
6009-10/2 were Bolton 9,10,12.
6016-6037 were Rochdale 16-37.
6038 was an ex-Daimler demonstrator delivered to Selnec in Rochdale livery as 38.
6040-6049 were ordered by Rochdale. They were delivered in Selnec livery.
6054-6057 were Bolton 14-17.
6059 was previously EX59, being renumbered in April 1973. It was renumbered 1703 later in 1973.
6060-6064 were Leigh 20-24.
6081 was Bury 81.
6082 was Ramsbottom 12. It was new to Halifax and then operated for Warrington before joining the Ramsbottom fleet in 1968.
6087-6097 were Bury 87-97.

6138-6254: Selnec Northern ex-Rochdale double deckers

6138-6152	JDK738-752	Daimler CVG6	Weymann	H33/26R	1953
6153-6167	KDK653-667	Daimler CVG6	Weymann	H33/26R	1954
6168-6197	NDK968-997	AEC Regent V D2RA6G	Weymann	H33/28R	1956
6198-6207	ODK698-707	AEC Regent V D2RA6G	Weymann	H33/28R	1956
6208-6218	RDK408-418	AEC Regent V D2RA	Weymann	H33/28R	1957-8
6219-6222	TDK319-322	AEC Regent V D2RA	Weymann	H33/28RD	1959
6223-6227	6323-6327DK	Daimler Fleetline CRG6LX	Weymann	H43/34F	1964
6228-6234	EDK128-134C	Daimler Fleetline CRG6LX	Metro-Cammell	H43/34F	1965
6235-6244	KDK135-144F	Daimler Fleetline CRG6LX	Northern Counties	H43/32F	1968
6245-6249	TNB747-751K	Daimler Fleetline CRG6LXB	Northern Counties	H43/32F	1972
6250-6254	TNB757-761K	Daimler Fleetline CRG6LXB	Northern Counties	H45/27D	1972

6138-6244 were Rochdale 238-344.
6245-6254 had chassis ordered by Rochdale. They were previously EX7-11, 17-21.
6168-6206 had Gardner 6LW engines; 6207 had been re-engined in 1963 with an AEC 9.6 litre engine.

143

63xx series: Selnec Northern ex-Bury double-deckers

6302-6316	REN102-116	Leyland Atlantean PDR1/1	Metro-Cammell	H41/33F	1963
6317-6331	TEN117-131	Daimler Fleetline CRG6LX	Alexander	H43/31F	1964
6332-6337	AEN832-837C	Daimler Fleetline CRG6LX	East Lancs	H43/31F	1965
6338-6343	HEN538-543F	Daimler Fleetline CRG6LX	East Lancs	H45/28F	1968
6344-6350	NEN504-510J	Daimler Fleetline CRG6LX	East Lancs	H45/28D	1970
6351-6375	GEN201-225	Leyland Titan PD3/6	Weymann	H41/32RD	1958-9
6376-6377	BEN176-177	AEC Regent III 9613A	Weymann	H30/26R	1952
6378-6384	BEN178-184	Leyland Titan PD2/12	Weymann	H30/26R	1953
6386	BEN186	Leyland Titan PD2/12	Weymann	H37/28F	1953
6387-6390	FEN587-590E	Leyland Titan PD2/37	East Lancs	H45/33F	1967
6391-6393	KEN231-233G	Leyland Atlantean PDR1A/1	East Lancs	H45/27D	1969
6395-6399	TNB752-756K	Daimler Fleetline CRG6LXB	Northern Counties	H43/32F	1972

6302-6343 were Bury 102-143.
6344-6350 were ordered by Bury. They were delivered in Selnec livery.
6351-84/6-93 were Bury 201-225, 176-184/6-90, 1-3.
6395-6399 had chassis ordered by Bury. They were previously EX12-16.

64xx: Selnec Northern ex-Ramsbottom double-deckers

6401	MTC998	Leyland Titan PD2/3	Leyland	H30/26R	1951
6402	247STD	Leyland Titan PD2/24	East Lancs	H35/28R	1961
6403	367XTE	Leyland Titan PD2A/30	East Lancs	H35/28R	1962
6404	9459TE	Leyland Titan PD2A/30	East Lancs	H35/28F	1963
6405	TTB879D	Leyland Titan PD3A/1	East Lancs	H41/32F	1965
6406-6407	DTC415-416E	Leyland Titan PD2/37	East Lancs	H41/32F	1966
6408-6409	FTF702-703F	Leyland Titan PD3/4	East Lancs	H41/32F	1967
6410	OTJ334G	Leyland Titan PD3/14	East Lancs	H41/32F	1969
6411	TTD386H	Leyland Titan PD3/14	East Lancs	H41/32F	1969

MTC998 was Ramsbottom 29.
6401-6410 were Ramsbottom 1-10.
6411 was ordered by Ramsbottom and was delivered to Selnec in Ramsbottom livery as 11.

6551-6816: Selnec Northern ex-Bolton double-deckers

	CWH703/7	Leyland Titan PD2/4	Leyland	H32/26R	1948
	DBN346	Leyland Titan PD2/4	Leyland	H32/26R	1949
6551-6560	GWH501-510	Leyland Titan PD2/13	Metro-Cammell	H31/27R	1955
	GWH511-512	Leyland Titan PD2/13	Metro-Cammell	H31/27R	1955
6563-6565	GWH513-515	Leyland Titan PD2/13	Metro-Cammell	H32/28R	1955
6566	JBN140	Leyland Titan PD2/12	Bond	H32/28R	1955
6567-6572	JBN143-148	Leyland Titan PD2/12	Bond	H33/27R	1955
	JBN149	Leyland Titan PD2/12	Metro-Cammell	H34/28R	1955
6574-6575	JBN150-151	Leyland Titan PD2/12	Metro-Cammell	H35/28RD	1956
6576-6584	JBN152-160	Daimler CVG6K	East Lancs	H34/28R	1957
6585-6594	KWH565-574	Daimler CVG6K	Metro-Cammell	H34/28F	1957
6595-6605	KWH575-585	Leyland Titan PD2/37	Metro-Cammell	H41/33R	1958
6606-6612	MBN161-167	Leyland Titan PD3/5	East Lancs	H41/33R	1958
6613-6622	MBN168-177	Leyland Titan PD2/37	Metro-Cammell	H34/28R	1959
6623-6627	NBN431-435	Leyland Titan PD3/4	Metro-Cammell	H41/32F	1959
6628-6632	NBN436-440	Leyland Titan PD2/27	East Lancs	H41/32F	1961
6633-6642	PBN651-660	Daimler CVG6-30	Metro-Cammell	FH35/27F	1960
6643-6650	PBN661-668	Daimler CVG6-30	East Lancs	H41/32F	1961
6651-6661	SBN751-761	Leyland Titan PD3/4	East Lancs	H41/32F	1961
6662-6667	SBN762-767	AEC Regent V 2D3RA	Metro-Cammell	H40/32F	1961
6668-6676	UBN901-909	Leyland Titan PD3A/2	East Lancs	FH41/32F	1962
6677-6684	UBN910-917	Leyland Titan PD3A/2	Metro-Cammell	FH41/31F	1962-63
6685-6692	UWH185-192	Leyland Atlantean PDR1/1	East Lancs	H45/38F	1963
6693-6699	UWH193-199	Leyland Atlantean PDR1/1	East Lancs	H43/35F	1964
6700-6703	ABN200-203B	Leyland Atlantean PDR1/1	Neepsend	H45/33F	1964
6704-6706	ABN204-206B	Leyland Atlantean PDR1/1	East Lancs	H45/33F	1964
6707-6710	ABN207-210B	Leyland Atlantean PDR1/1	Neepsend	H45/33F	1964
6711	ABN211B	Leyland Atlantean PDR1/1	East Lancs	H45/33F	1965
6712-6718	ABN212-218C	Leyland Atlantean PDR1/1	Metro-Cammell	H45/33F	1965
6719-6726	ABN219-226C	Leyland Atlantean PDR1/1	East Lancs	H45/33F	1965
6727-6734	FBN227-234C	Leyland Atlantean PDR1/1	East Lancs	H45/33F	1966
6735	FBN235D	Leyland Atlantean PDR1/1	East Lancs	H45/33F	1966
6736	FBN236C	Leyland Atlantean PDR1/1	East Lancs	H45/33F	1965
6737	FBN237D	Leyland Atlantean PDR1/1	East Lancs	H45/33F	1966
6738	FBN238C	Leyland Atlantean PDR1/1	East Lancs	H45/33F	1965
6739	FBN239D	Leyland Atlantean PDR1/1	East Lancs	H45/33F	1966
6740-6741	FBN240-241C	Leyland Atlantean PDR1/1	East Lancs	H45/33F	1965
6742-6756	GBN242-256C	Leyland Atlantean PDR1/1	East Lancs	H45/33F	1966
6757-6771	HWH257-271F	Leyland Atlantean PDR1/1	East Lancs	H43/33F	1967
6772-6786	MWH272-286G	Leyland Atlantean PDR1A/1	East Lancs	H45/33F	1968
6787-6801	OBN287-301H	Leyland Atlantean PDR1A/1	East Lancs	H43/29D	1969-70
6802-6816	TWH802-816K	Leyland Atlantean PDR2/1	East Lancs	H49/37F	1971-72

CWH703/7, DBN346, 6551-60, GWH511/2, 6563-72, JBN149, 6574-6787, 6791/5, 6800 were Bolton 353/7, 443, 51-287, 291/5, 300.
6788-90/2-4/6-9, 6801 were ordered by Bolton and were delivered to Selnec in Bolton livery as 288-290/2-4/6-9, 301.
6802-6816 were ordered by Bolton. They were delivered to Selnec in Selnec livery.

69xx series: Selnec Northern ex-Leigh double-deckers

6903	778YTB	Leyland Titan PD3A/3	East Lancs	L34/32R	1962
6905	ATB246D	AEC Renown 3B3RA	East Lancs	H41/31F	1966
6908	HTJ761B	AEC Renown 3B3RA	East Lancs	H41/31F	1964
6911/13	PTC112-113C	AEC Renown 3B3RA	Lydney	H41/31F	1965
6914	KTD766	Leyland Titan PD2/1	East Lancs	L27/26R	1950
6915	PTC114C	AEC Renown 3B3RA	Lydney	H41/31F	1965
6916-6918	KTD768-770	Leyland Titan PD2/1	East Lancs	L27/26R	1950
6925-6928	1972-1975TJ	AEC Renown 3B3RA	East Lancs	H41/31R	1963
6929	ATE190E	AEC Renown 3B3RA	East Lancs	H41/31F	1967
6930	HTJ762B	AEC Renown 3B3RA	East Lancs	H41/31F	1964
6931-6932	YTJ627-628D	AEC Renown 3B3RA	East Lancs	H41/31F	1966
6934	HTJ763B	AEC Renown 3B3RA	East Lancs	H41/31F	1964
6935	PTC115C	AEC Renown 3B3RA	East Lancs	H41/31F	1965
6936	ATB245D	AEC Renown 3B3RA	East Lancs	H41/31F	1966
6937	779YTB	Leyland Titan PD3A/3	East Lancs	L34/32R	1962
6938	HTD328B	AEC Renown 3B3RA	East Lancs	H41/31F	1964
6939	HTJ764B	AEC Renown 3B3RA	East Lancs	H41/31F	1964
6947-6951	NTE381-387	AEC Regent III 9613E	East Lancs	L27/26R	1952
6952-6954	WTE21-25	Leyland Titan PD2/20	East Lancs	L30/28R	1955
6955-6959	722-724ATE	Leyland Titan PD2/20	East Lancs	L30/28R	1957
6960-6961	491-495DTC	Leyland Titan PD2/30	East Lancs	L30/28R	1957-58
6962-6963	223-224FTC	Dennis Loline 1 6LW	East Lancs	H41/33R	1958
6964-6965	878-879GTF	Dennis Loline 1 6LW	East Lancs	H41/31R	1959
	267-268WTE	Dennis Loline III 6LX	East Lancs	H41/31R	1961

6903/5/8, 6911/3-8, 6925-32/4-65 were Leigh 3, 5, 8, 11, 13-18, 25-32, 34-65.

7000 series: Standard double-deckers

Fleet no.	Registration	Chassis	Body	Layout	Year
7001-7071	VNB101-171L	Leyland Atlantean AN68/1R	Park Royal	H43/32F	1972-73
7072-7109	WBN950-987L	Leyland Atlantean AN68/1R	Park Royal	H43/32F	1972-73
7110-7145	XJA501-536L	Leyland Atlantean AN68/1R	Park Royal	H43/32F	1972-73
7146-7150	VNB172-176L	Leyland Atlantean AN68/1R	Northern Counties	H43/32F	1972
7151-7162	WBN988-999L	Daimler Fleetline CRG6LXB	Park Royal	H43/32F	1973
7163-7186	WWH21-44L	Daimler Fleetline CRG6LXB	Park Royal	H43/32F	1973
7187-7205	XJA537-555L	Daimler Fleetline CRG6LXB	Park Royal	H45/27D	1973
7206-7251	VNB177-222L	Daimler Fleetline CRG6LXB	Northern Counties	H43/32F	1972-73
7252-7270	VNB223-241L	Daimler Fleetline CRG6LXB	Northern Counties	H43/32F	1973
7271-7279	YNA271-279M	Daimler Fleetline CRG6LXB	Northern Counties	H45/27D	1973
7280-7281	WWH45-46L	Daimler Fleetline CRG6LXB	Northern Counties	H43/32F	1972
7282-7299	WWH47-64L	Daimler Fleetline CRG6LXB	Northern Counties	H43/32F	1973
7300-7324	XJA556-580L	Daimler Fleetline CRG6LXB	Northern Counties	H43/32F	1973
7325-7415	YNA280-370M	Daimler Fleetline CRG6LXB	Northern Counties	H43/32F	1973-74
7416-7421	BNE732-737N	Daimler Fleetline CRG6LXB	Northern Counties	H43/32F	1974
7422-7423	GNC287-288N	Daimler Fleetline CRG6LXB	Northern Counties	H43/32F	1974
7424-7426	BNE740-742N	Daimler Fleetline CRG6LXB	Northern Counties	H43/32F	1974
7427	GNC289N	Daimler Fleetline CRG6LXB	Northern Counties	H43/32F	1974
7428	GNC294N	Daimler Fleetline CRG6LXB	Northern Counties	H43/32F	1974
7429-7433	GND489-493N	Daimler Fleetline CRG6LXB	Northern Counties	H43/32F	1974
7434-7442	GND500-508N	Daimler Fleetline CRG6LXB	Northern Counties	H43/32F	1974
7443-7447	GDB162-166N	Daimler Fleetline CRG6LXB	Northern Counties	H43/32F	1974
7448-7452	HJA116-120N	Daimler Fleetline CRG6LXB	Northern Counties	H43/32F	1975
7453-7455	JND981-983N	Daimler Fleetline CRG6LXB	Northern Counties	H43/32F	1975
7456-7459	JDB108-111N	Daimler Fleetline CRG6LXB	Northern Counties	H43/32F	1975
7460-7464	JVM991-995N	Daimler Fleetline CRG6LXB	Northern Counties	H43/32F	1975
7465-7469	KBU906-910P	Daimler Fleetline CRG6LXB	Northern Counties	H43/32F	1975
7470-7484	LJA470-484P	Daimler Fleetline CRG6LXB	Northern Counties	H43/32F	1975-76
7485-7500	PRJ485-500R	Daimler Fleetline CRG6LXB	Northern Counties	H43/32F	1976
7501-7508	BNE751-758N	Leyland Atlantean AN68/1R	Northern Counties	H43/32F	1974
7509-7512	GNC290-293N	Leyland Atlantean AN68/1R	Northern Counties	H43/32F	1974
7513-7518	GND494-499N	Leyland Atlantean AN68/1R	Northern Counties	H43/32F	1974
7519	GND510N	Leyland Atlantean AN68/1R	Northern Counties	H43/32F	1974
7520-7534	GDB167-181N	Leyland Atlantean AN68/1R	Northern Counties	H43/32F	1974-75
7535-7555	HNB25-45N	Leyland Atlantean AN68/1R	Northern Counties	H43/32F	1975
7556-7559	HJA112-115N	Leyland Atlantean AN68/1R	Northern Counties	H43/32F	1975
7560-7566	JND984-990N	Leyland Atlantean AN68A/1R	Northern Counties	H43/32F	1975
7567-7570	JDB112-115N	Leyland Atlantean AN68A/1R	Northern Counties	H43/32F	1975
7571	KBU918P	Leyland Atlantean AN68A/1R	Northern Counties	H43/32F	1975-76
7572-7577	JDB117-122N	Leyland Atlantean AN68A/1R	Northern Counties	H43/32F	1976-77
7578-7579	JVM989-990N	Leyland Atlantean AN68A/1R	Northern Counties	H43/32F	1977
7580-7586	KBU911-917P	Leyland Atlantean AN68A/1R	Northern Counties	H43/32F	1977
7587-7596	KDB680-689P	Leyland Atlantean AN68A/1R	Northern Counties	H43/32F	1977
7597-7599	LNA250-252P	Leyland Atlantean AN68A/1R	Northern Counties	H43/32F	1977-78
7600-7652	LJA600-652P	Leyland Atlantean AN68A/1R	Park Royal	H43/32F	1978
7653-7700	ONF653-700R	Leyland Atlantean AN68A/1R	Park Royal	H43/32F	1978-79
7701-7730	RJA701-730R	Leyland Atlantean AN68A/1R	Park Royal	H43/32F	1979
7731-7760	SRJ731-760R	Leyland Atlantean AN68A/1R	Northern Counties	H43/32F	1978
7761-7800	UNA761-800S	Leyland Atlantean AN68A/1R	Northern Counties	H43/32F	1978-79
7801-7816	RJA801-816R	Leyland Atlantean AN68A/1R	Northern Counties	H43/32F	1979
7817-7872	UNA817-872S	Leyland Atlantean AN68A/1R	Northern Counties	H43/32F	1979-80
7873-7902	WVM873-902S	Leyland Fleetline FE30AGR	Northern Counties	H43/32F	1980
7903-7932	ANC903-932T	Leyland Fleetline FE30AGR	Northern Counties	H43/32F	1980
7933-7960	BNC933-960T	Leyland Fleetline FE30AGR	Northern Counties	H43/32F	1980
7961-8000	not used				
8001-8020	XBU1-20S	Leyland Fleetline FE30AGR	Northern Counties	H43/32F	1978
8021-8050	ANA21-50T	Leyland Fleetline FE30AGR	Northern Counties	H43/32F	1978-79
8051-8100	BVR51-100T	Leyland Fleetline FE30AGR	Northern Counties	H43/32F	1979
8101-8125	HDB101-125V	Leyland Fleetline FE30AGR	Northern Counties	H43/32F	1979-80
8126-8140	KDB126-140V	Leyland Fleetline FE30AGR	Northern Counties	H43/32F	1980
8141-8142	GNF16-17V	Leyland Fleetline FE30AGR	Northern Counties	H43/32F	1980
8143-8150	MNC486-493W	Leyland Fleetline FE30AGR	Northern Counties	H43/32F	1980
8151-8200	VBA151-200S	Leyland Atlantean AN68A/1R	Northern Counties	H43/32F	1978
8201-8206	XRJ201-206S	Leyland Atlantean AN68A/1R	Northern Counties	H43/32F	1978
8207-8239	ANA207-239T	Leyland Atlantean AN68A/1R	Northern Counties	H43/32F	1978-79
8240	FVR240V	Leyland Atlantean AN68A/1R	Northern Counties	H43/32F	1979
8241	ANA241T	Leyland Atlantean AN68A/1R	Northern Counties	H43/32F	1979
8242-8300	FVR242-300V	Leyland Atlantean AN68A/1R	Northern Counties	H43/32F	1979-80
8301-8303	KDB301-303V	Leyland Atlantean AN68A/1R	Northern Counties	H43/32F	1980
8304-8350	MNC504-550W	Leyland Atlantean AN68A/1R	Northern Counties	H43/32F	1980-81
8351-8400	ORJ351-400W	Leyland Atlantean AN68A/1R	Northern Counties	H43/32F	1981
8401-8411	MRJ401-411W	Leyland Atlantean AN68A/1R	Northern Counties	H43/32F	1981
8412-8525	SND412-525X	Leyland Atlantean AN68A/1R*	Northern Counties	H43/32F	1981-82
8526-8530	SND526-530X	Leyland Atlantean AN68D/1R	Northern Counties	H43/32F	1982
8531-8655	ANA531-655Y	Leyland Atlantean AN68D/1R	Northern Counties	H43/32F	1982-83
8656-8700	A656-700HNB	Leyland Atlantean AN68D/1R	Northern Counties	H43/32F	1983-84
8701-8730	A701-730LNC	Leyland Atlantean AN68D/1R	Northern Counties	H43/32F	1984
8731-8765	A731-765NNA	Leyland Atlantean AN68D/1R	Northern Counties	H43/32F	1984
8766-8825	Atlantean order cancelled and converted to Olympians.				

* 8425, 8448/9, 8455/6, 8460-6/8-72/4-8, 8480/1/3-5, 8490/3/4, 8500-8525 are AN68B/1R.

The following registrations were booked but not used because of delivery delays:

Fleet no.	Intended registration	Actual registration
7271-7279	VNB242-250L	YNA271-279M
7325-7328	XJA581-584L	YNA280-283M
7416-7445	YNA371-400M)	BNE732-7N, GNC287/8N, BNE740-2N, GNC289/94N,
7416-7434	BNE732-750M)	GND489-93, 500-8N, GDB162-4N.
7465	JVM996N	KBU906P
7485-7500	LJA485-500P	PRJ485-500R
7571	JDB116N	KBU918P
7653-7675	LJA653-675P	ONF653-675R
7817-7850	RJA817-850R	UNA817-850S
8207-8220	XRJ207-220S	ANA207-220T
8304-8315	KDB304-315V	MNC504-515W
8412-8420	MRJ412-420W	SND412-420X

Minibuses

1240	C240JTU	Iveco Ford 49.10	Robin Hood	B21F	1986
1500	C240JTU	Iveco Ford 49.10	Robin Hood	B21F	1986
1501-1550	D501-550MJA	Iveco Ford 49.10	Robin Hood	B21F	1987
E181-189	E181-189CNE	Iveco Ford 49.10	Northern Counties	B22F	1987

1240 was an Iveco Ford demonstrator. It was purchased in 1987 and renumbered 1500.
E181-189 were operated on loan from Northern Counties from June 1988. They were painted overall white.

1601-1650	D601-650MDB	MCW Metrorider	MCW	B23F	1987
1651-1680	D651-680NNE	MCW Metrorider	MCW	B23F	1987
1681	D474PON	MCW Metrorider	MCW	B23F	1987
1682/3	E148/50KYW	MCW Metrorider	MCW	B25F	1987
1684-6	E638/7/6 KYW	MCW Metrorider	MCW	B25F	1987
1687	E929KYR	MCW Metrorider	MCW	B25F	1987

1639 was fitted with 19 coach seats in 1990 and renumbered 21 in the coach series.
1681-7 were acquired from London Buses in 1993 where they were MR 14, 48, 50, 62, 61, 60, 53.

1771-1782	D771-782RBU	Dodge S56	Northern Counties	B20F	1987
1783-1800	E783-800SJA	Dodge S56	Northern Counties	B20F	1987
1801-1840	C801-840CBU	Dodge S56	Northern Counties	B18F	1986
1841-1865	D841-865LND	Dodge S56	Northern Counties	B20F	1987
1866-1905	D866-905MDB	Dodge S56	Northern Counties	B20F	1987
1906-1955	D906-955NDB	Dodge S56	Northern Counties	B20F	1987
1956-1980	D956-980PJA	Dodge S56	Northern Counties	B20F	1987
1981-1999	E981-999SJA	Dodge S56	Northern Counties	B20F	1987
2000	E200SVR	Dodge S56	Northern Counties	B20F	1987

Selnec Cheshire

1-6	YJA1-6	Daimler Fleetline CRG6LX	Alexander	H44/31F	1963
8	YJA8	Daimler Fleetline CRG6LX	Alexander	H44/31F	1963
10-13	YJA10-13	Daimler Fleetline CRG6LX	Alexander	H44/31F	1963
15-21	YJA15-21	Daimler Fleetline CRG6LX	Alexander	H44/31F	1963
100-114	AJA100-114B	Daimler Fleetline CRG6LX	Alexander	H44/31F	1964
115-129	AJA115-129B	AEC Renown 3B3RA	Park Royal	H42/30F	1964
165-189	DDB165-189C	Daimler Fleetline CRG6LX	Alexander	H44/31F	1965
197-210	FJA197-210D	Daimler Fleetline CRG6LX	Alexander	H44/31F	1966
270-275	KJA270-275F	Bristol RESL6G	Marshall	B45F	1968
277-290	KJA277-290F	Bristol RESL6G	Marshall	B45F	1968
291-301	KJA291-301G	Bristol RESL6G	Marshall	B45F	1968
315-344	NJA315-344H	Bristol RELL6G	Alexander	B49F	1970
400-424	AJA400-424L	Bristol VRTSL6G	ECW	H43/32F	1973
721-724	LDB721-724	AEC Reliance MU3RA	Weymann	B43F	1957
727/9	LDB727/9	AEC Reliance MU3RA	Weymann	B43F	1957
783-796	LDB783-796	Leyland Tiger Cub PSUC1/1	Willowbrook	DP43F	1960
812-814	RDB812-814	Dennis Loline I 6LX	East Lancs	H39/32F	1960
815-826	RDB815-826	Dennis Loline I 0.600	East Lancs	H39/32F	1960
832/3	RDB832/3	AEC Reliance 2MU3RA	Alexander	DP41F	1961
838-841	RDB838-841	AEC Reliance 2MU3RA	Alexander	DP41F	1961
845-851	RDB845-851	AEC Reliance 2MU3RA	Alexander	DP41F	1961
852-871	RDB852-871	AEC Reliance 2MU3RA	Willowbrook	B43F	1961
872-895	RDB872-895	Dennis Loline III 6LX	Alexander	H39/32F	1961
907	VDB907	Leyland Leopard PSU3/3RT	Alexander	DP49F	1962
909-916	VDB909-916	Leyland Leopard PSU3/3RT	Alexander	DP49F	1962
925-929	VDB925-929	AEC Reliance 2U3RA	Willowbrook	B53F	1963
932-933	VDB932-933	AEC Reliance 2U3RA	Willowbrook	DP51F	1963
937-945	VDB937-945	AEC Reliance 2U3RA	Willowbrook	DP51F	1963
952-953	VDB952-953	Leyland Leopard PSU3/3RT	Alexander Y	DP49F	1963
958-960	VDB958-960	Leyland Leopard PSU3/3RT	Alexander Y	DP49F	1963
970-971	VDB970-971	AEC Renown 3B3RA	Park Royal	H42/32F	1963

All of the above were ex-North Western Road Car (same fleet numbers) except 400-424, which were ordered by North Western but delivered to Selnec.

To leave 100-series numbers free for new Leyland Nationals ex North Western buses between 100 and 210 were renumbered in 1975 as follows:

100-114	4100-4114
115-129	1915-1929
165-189	4165-4189
197-210	4197-4210

At the same time 400-424 were renumbered 1400-1424 in a new non-standard double-deck series.

Routemasters

2200	CUV200C	AEC Routemaster	Park Royal	H36/28R	1965
2201	136CLT	AEC Routemaster	Park Royal	H36/28R	1962
2202	776DYE	AEC Routemaster	Park Royal	H36/28R	1963
2203	604DYE	AEC Routemaster	Park Royal	H36/28R	1963
2204	CUV162C	AEC Routemaster	Park Royal	H36/28R	1965
2205	618DYE	AEC Routemaster	Park Royal	H36/28R	1963
2206-7	WLT378,429	AEC Routemaster	Park Royal	H36/28R	1960
2208	WLT698	AEC Routemaster	Park Royal	H36/28R	1961
2209	807DYE	AEC Routemaster	Park Royal	H36/28R	1964

2200-2209 were acquired from London Buses in 1988 where they were RM 2200, 1136, 1776, 1604, 2162, 1618, 378, 429, 698, 1807.

16xx and 3200 series: ex-Wigan

1673	JJP501	Leyland Tiger Cub PSUC1/1	Massey	B43F	1962
1675	DJP468E	Leyland Panther Cub PRSC1/1	Massey	B43D	1967
1676	EEK1F	Leyland Panther Cub PRSC1/1	Massey	B43D	1967
1680-1688	HJP950-958H	Leyland Panther PSUR1A/1	Northern Counties	B46D	1970
1690-1691	HJP960-961H	Leyland Panther PSUR1A/1	Northern Counties	B46D	1970
3200	DJP754	Leyland Titan PD2/30	Massey	H33/28R	1958
3201-3204	EJP501-504	Leyland Titan PD3/2	Massey	H41/31F	1959
3205-3210	EJP505-510	Leyland Titan PD3/2	Massey	H39/30F	1959
3211-3213	GJP8-10	Leyland Titan PD3/2	Massey	H41/29F	1960
3214-3219	GJP11-16	Leyland Titan PD3/2	Northern Counties	H40/30F	1960
3220-3222	GJP17-19	Leyland Titan PD3/2	Massey	H41/29F	1960
3230-3232	HEK705-707	Leyland Titan PD3A/2	Massey	H41/29F	1961
3233-3236	HJP1-4	Leyland Titan PD3A/2	Massey	H41/29F	1962
3237-3243	HJP5-11	Leyland Titan PD3A/2	Northern Counties	H40/30F	1962
3244-3247	JJP502-505	Leyland Titan PD2A/27	Massey	H37/27F	1962
3248-3251	JUP506-509	Leyland Titan PD2A/27	Massey	H36/28F	1962
3255-3260	KEK739-744	Leyland Titan PD2A/27	Massey	H37/27F	1963
3261-3266	KEK745-750	Leyland Titan PD2A/27	Massey	H37/27F	1963
3270-3279	AEK1-10B	Leyland Titan PD2/37	Massey	H37/27F	1964
3280-3281	DEK2-3D	Leyland Titan PD2/37	Massey	H37/27F	1966
3285-3288	DEK4-7E	Leyland Titan PD2/37	Northern Counties	H37/27F	1967
3290-3298	FEK1-9F	Leyland Titan PD2/37	Massey	H37/27F	1968
3300	FJP566G	Leyland Atlantean PDR1A/1	Northern Counties	H44/27D	1968
3301-3309	GJP2-10G	Leyland Atlantean PDR1A/1	Northern Counties	H47/29F	1969
3310-3321	KJP20-31J	Leyland Atlantean PDR2/1	Northern Counties	H48/31D	1971
3330-3339	NEK1-10K	Leyland Atlantean AN68/2R	Northern Counties	H48/31D	1972

1673/5/6 were Wigan 21, 20, 22.
1680-1688/90/91 were Wigan 80-88, 90/1.
3200 was a non-psv training bus, ex Wigan 115.
3201-3222 were Wigan 63, 60-62, 64-68, 128, 136-138, 120-122, 129, 134/5, 141, 143/4.
3230-3251 were Wigan 57-59, 39, 40, 42, 49, 51, 54, 70/1, 74, 77, 79, 35-37, 145, 130-133.
3255-3266 were Wigan 41, 43-45, 47/8, 52, 55/6, 69, 75/6.
3270-3285 were Wigan 146-150, 30, 11, 24, 29, 31, 139, 140, 72/3, 78, 113.
3290-3298 were Wigan 25-28, 32-34, 38, 46.
3300-3321 were Wigan 166, 157-165, 92-97, 151-156.
3300-3309 were Wigan 1-10.

The experimental vehicles

Original no	Renumbered					
EX1-6	5466-5471	PNF941-946J	Leyland Atlantean PDR1A/1	Northern Counties	H43/32F	1971
EX7-11	6245-6249	TNB747-751K	Daimler Fleetline CRG6LXB	Northern Counties	H43/32F	1972
EX12-16	6395-6399	TNB752-756K	Daimler Fleetline CRG6LXB	Northern Counties	H43/32F	1972
EX17-21	6250-6254	TNB757-761K	Daimler Fleetline CRG6LXB	Northern Counties	H45/27D	1972
EX22-29	not used		Leyland National 1151/2R		B46D	1972
EX30-37	1330-1337	TXJ507-514K	Leyland National 1051/2R		B40D	1973
EX38-41	1338-1341	VVM601-604L	Metro-Scania BR111MH		B44D	1972
EX42-49	1342-1349	TXJ515-522K	Metro-Scania BR110MH		B43D	1973
EX50-53	1350-1353	VVM605-608L	Mercedes-Benz O.305	Northern Counties	B43D	1973
EX54-55	1354-1355	WVM668-669L	Seddon Pennine IV:236	Seddon	DP25F	1972
EX56-58	1700-1702	YDB453-455L	Seddon Pennine IV:236	Seddon	DP25F	1972
EX59	6059	XBN976L	Metro-Scania BR110		B40D	1973
EX60	1360	VVM609L	Seddon-Chloride	Seddon	B43D	1973
EX61	1361	XVU387M	Seddon-Lucas	Seddon	B19F	1975
EX62	1362	GNC276N	Leyland-Crompton	Willowbrook	B19F	1972

EX1-6 had chassis ordered by Ashton and were the first prototypes of Selnec's standard double-decker.
EX7-11, 17-21 had chassis ordered by Rochdale.
EX12-16 had chassis ordered by Bury.
EX30 was the first production Leyland National.
EX59 was subsequently renumbered 1703.
EX60 had an encapsulated engine and was marketed as the Hush Bus.
EX61 was delivered as XVU87M. It was promoted as the Silent Rider.
EX100 was on loan from the Department of Trade and Industry during 1973.
EX61 and EX62 did not carry their new numbers 1361/2.

1300 series: non-standard single-deckers

1300-1318	XVU367-385M	Leyland National 1051/1R		B41F	1974
1319	GND511N	Leyland National 1051/1R		B41F	1974
1320-1325	BNE763-768N	Bristol LH6L	ECW	B43F	1974
1330-1337	TXJ507-514K	Leyland National 1151/2R		B40D	1972
1338-1341	VVM601-604L	Leyland National 1051/2R		B45D	1972
1342-1349	TXJ515-522K	Metro-Scania BR111MH		B41D	1973
1350-1353	VVM605-608L	Metro-Scania BR110MH		B43D	1973
1354-1355	WVM668-669L	Mercedes-Benz O.305	Northern Counties		
1356-1359	not used (see 1700-1703)				
1360	VVM609L	Metro-Scania BR110		B40D	1973
1361	XVU387M	Seddon-Chloride	Seddon	B43D	1973
1362	GNC276N	Seddon-Lucas	Seddon	B19F	1975

1319 was to have been registered XVU386M then BNE731N.
1320-1325 were ordered by Wigan. They were delivered in GMT livery. They were originally to have been registered XVU388-393M.
1330-1355/60-62 were previously EX30-55, 60-62.

1400 series: non-standard double-deckers

1400-1424	AJA400-424L	Bristol VRTSL6G	ECW	H43/32F	1973
1425-1434	GNC277-286N	MCW Metropolitan BR111DH	Metro-Cammell	H44/29F	1974
1435	LNA258P	Foden-NC	Northern Counties	H43/32F	1976
1436	PNE358R	Foden-NC	Northern Counties	H43/32F	1976
1437-1438	HDB437-438V	Dennis Dominator	Northern Counties	H43/32F	1980
1439-1440	TND439-440X	Dennis Dominator	Northern Counties	H43/32F	1981
1441-1442	not used				
1443-1445	B443-445TVU	allocated to Volvo Citybuses delivered as 1481-1483.			
1446	NNA134W	Volvo Ailsa MkII	Northern Counties	H44/35F	1980
1447-1448	WRJ447-448X	Volvo Ailsa MkII	Northern Counties	H44/35F	1982
1449-1450	allocated to Volvo Ailsas which were cancelled.				
1451	NJA568W	Leyland Olympian TL11	Northern Counties	H43/30F	1980
1461-1462	FWH461-462Y	Scania BR112DH	Northern Counties	H43/32F	1983
1463-1467	H463-467GVM	Scania N113DR	Northern Counties	H47/28F	1991
1471-1473	A471-473HNC	Dennis Falcon V	Northern Counties	H47/37F	1984
1481-1483	C481-483CBU	Volvo Citybus	Northern Counties	H46/33F	1986

1400-1424 were previously 400-424. They were ordered by North Western.
1425-1434 were to have been BNE759-768N.

Lancashire United Transport: Fleet acquired by GMT

LUT fleet no.	GMT fleet no.					
18-23		113-118JTD	Guy Arab IV 6LW	N Counties	H42/32R	1959
25		120JTD	Guy Arab IV 6LW	N Counties	H42/32R	1959
27		122JTD	Guy Arab IV 6LX	N Counties*	H42/32R	1959
40		531RTB	Guy Arab IV 6LW	N Counties	H41/32F	1960
41		532RTB	Guy Arab IV 6LW	N Counties	H42/32R	1961
43-44		534-535RTB	Guy Arab IV 6LW	N Counties	H42/32R	1960
50-57		141-148NTF	Guy Arab IV 6LW	Metro-Cammell	H41/32F	1961
61-65		501-505VTB	Guy Arab IV 6LW	N Counties	H41/32F	1961
67-68		507-508VTB	Guy Arab IV 6LW	N Counties	H41/32F	1961
70-71		510-511VTB	Guy Arab IV 6LW	N Counties	H41/32F	1961
73/5		513/5VTB	Guy Arab IV 6LW	N Counties	H41/32F	1961
77-80		517-520VTB	Guy Arab IV 6LW	N Counties	H41/32F	1961
97-99	2305-2307	561-563TD	Daimler Fleetline CRG6LX	N Counties	H43/33F	1962
100-102		564-566TD	Daimler Fleetline CRG6LX	N Counties	H43/33F	1962
103-119		567-583TD	Daimler Fleetline CRG6LX	N Counties	H43/31F	1962
120-135		6204-6219TF	Guy Arab IV 6LW	N Counties	H41/32F	1963
136		6220TF	Guy Arab V 6LW	N Counties	H41/32F	1963
138		4611TF	Daimler Fleetline CRG6LX	N Counties	H43/31F	1963
139	2308	4612TF	Daimler Fleetline CRG6LX	N Counties	H43/31F	1963
140		4613TF	Daimler Fleetline CRG6LX	N Counties	H43/31F	1963
141	2309	4614TF	Daimler Fleetline CRG6LX	N Counties	H43/31F	1963
142		4615TF	Daimler Fleetline CRG6LX	N Counties	H43/31F	1963
143-144		8087-8088TE	AEC Reliance	Plaxton	B50F	1963
151-155		DTF581-585B	Leyland Leopard PSU3/1	Plaxton	DP49F	1964
156-158		DTF586-588B	Leyland Leopard PSU3/1	Plaxton	B50F	1964
159-160		KTC792-793C	Guy Arab V 6LX	N Counties	H41/32F	1965
161		JTD299B	Guy Arab V 6LX	N Counties	H41/32F	1964
162-163		KTC794-795C	Guy Arab V 6LX	N Counties	H41/32F	1965
164		KTC791C	Guy Arab V 6LX	N Counties	H41/32F	1965
165	2451	HTJ521B	Guy Arab V 6LX	N Counties	H41/32F	1964
166	2452	JTD300B	Guy Arab V 6LX	N Counties	H41/32F	1964
167	2453	HTJ522B	Guy Arab V 6LX	N Counties	H41/32F	1964
168-170		HTJ523-525B	Daimler Fleetline CRG6LX	N Counties	H43/31F	1964
171-175		ETD941-945B	Daimler Fleetline CRG6LX	N Counties	H43/31F	1964
176	2310	ETD946B	Daimler Fleetline CRG6LX	N Counties	H43/31F	1964
177-179		ETD947-949B	Daimler Fleetline CRG6LX	N Counties	H43/31F	1964
180	2311	ETD950B	Daimler Fleetline CRG6LX	N Counties	H43/31F	1964
181-185		PTE631-635C	Daimler Fleetline CRG6LX	N Counties	H43/31F	1965
186-195		RTC351-360C	Guy Arab V 6LX	N Counties	H41/32F	1965
196-198		LTB305-307C	Leyland Leopard L2	Plaxton	C43F	1965
199-203		LTE264-268C	Leyland Leopard L2	Plaxton	C43F	1965
204-207		LTB301-304C	Leyland Leopard PSU3	Marshall	B50F	1965
208-210		LTE261-263C	Leyland Leopard PSU3	Marshall	B50F	1965
211-213		TTF175-177D	Leyland Tiger Cub PSUC1/11	Duple	B44F	1966
214-217		UTC766-769D	Leyland Leopard L2T	Plaxton	C43F	1966
218-219		WTE141-142D	Guy Arab V 6LX	N Counties	H41/32F	1966
220	2454	WTE143D	Guy Arab V 6LX	N Counties	H41/32F	1966
221	2455	WTE144D	Guy Arab V 6LX	N Counties	H41/32F	1966
222		WTE145D	Guy Arab V 6LX	N Counties	H41/32F	1966
223-231		WTE146-154D	Guy Arab V 6LX	N Counties	H41/32F	1966
232-233	2456-2457	WTE155-156D	Guy Arab V 6LX	N Counties	H41/32F	1966
234-235		WTE157-158D	Guy Arab V 6LX	N Counties	H41/32F	1966
236	2458	WTE159D	Guy Arab V 6LX	N Counties	B49F	1966
237-240		WTE160-163D	Guy Arab V 6LX	N Counties	H41/32F	1966
241		YTC249D	Leyland Tiger Cub PSUC1/11	N Counties	B40D	1966
242-244		ETJ125-127F	Leyland Tiger Cub PSUC1/11	N Counties	B40D	1967
245-264		CTE471-490E	Bristol RESL6G	Marshall	B42D	1967
265-290		ETJ901-926F	Guy Arab V 6LX	Plaxton	H41/32F	1967
294-313		NTC111-130G	Bristol RESL6G	Alexander	B42D	1969

1700 series: small buses

1700-1702	YDB453-455L	Seddon Pennine IV:236	Seddon	DP25F	1972
1703	XBN976L	Seddon Pennine IV:236	Seddon	DP25F	1972
1704-1714	XVU334-344M	Seddon Pennine IV:236	Seddon	B23F	1973
1715-1734	XVU345-364M	Seddon Pennine IV:236	Seddon	B19F	1974
1735	BNE730M	Seddon Pennine IV:236	Seddon	B19F	1974
1736	GND509N	Seddon Pennine IV:236	Seddon	B19F	1975
1737-1742	HJA121-126N	Seddon Pennine IV:236	Seddon	B19F	1975
1743-1744	CNF1-2T	Ford Transit	Reeve Burgess	B17F	1979
1745-1746	AJA3-4T	Bedford CF	Reeve Burgess	B17F	1978
1747	XRJ9S	Bedford CF	Reeve Burgess	B17F	1978
1748	WNE10S	Bedford CF	Reeve Burgess	B17F	1978
1751-1770	C751-770YBA	Dennis Domino	Northern Counties	B24F	1985-86
1697	TWH69T	Daimler Fleetline FE30AGR	Northern Counties	B28F	1978
1701-1702	B701-702UVR	Leyland Cub CU435	Reeve Burgess	B30FL	1985
1721-1725	H721-725CNC	Renault S75	Northern Counties	B17FL	1990
1726-1728	H726-728FNC	Renault S75	Northern Counties	B17FL	1991

1700-1702 (Seddons) were previously 6056-58.
1703 was previously 6059 and originally EX59.
1734-1736 were to have been registered XVU364-366M.
1743-1748 were previously D1-4,9,10 in the Sale Dial-a-Ride fleet.
1697 was rebuilt in 1986 from double-deck 6938 and was originally LUT 522.

100 series: Leyland Nationals

101-105	HNB20-24N	Leyland National 10351/1R		B41F	1975
106-108	HNB46-48N	Leyland National 10351/1R		B41F	1975
109-115	HNE633-639N	Leyland National 10351/1R		B41F	1975
116-122	HNE646-652N	Leyland National 10351/1R		B41F	1975
123-127	HJA127-131N	Leyland National 10351/1R		B41F	1975
128-147	JNA584-603N	Leyland National 10351/1R		B41F	1975
148	JND991N	Leyland National 10351/1R		B41F	1975
149-150	JND998-999N	Leyland National 10351/1R		B41F	1975
151	JDB103N	Leyland National 10351/1R		B41F	1975
152-155	KBU889-892P	Leyland National 10351/1R		B41F	1975
156-157	JVM980-981P	Leyland National 10351/1R		B41F	1975
158-170	KBU893-905P	Leyland National 10351/1R		B49F	1977
171-185	RBU171-185R	Leyland National 11351A/1R		B49F	1977
186-205	ABA11-30T	Leyland National 11351A/1R		B49F	1979
221-227	NEN952-966R	Leyland National 11351A/1R		B49F	1977
228-250	PTD667-673S	Leyland National 11351A/1R		B49F	1978
	WBN462-484T	Leyland National 11351A/1R		B49F	1979

152-155 and 158-164 were to have been JDB104-107N and JVM982-988N.
206-250 were ex-LUT 465-479, 530-536, 543-565.

148

LUT fleet no.	GMT fleet no.				
314-317		RTF847-850G	Leyland Leopard PSU4/3	Plaxton	C43F	1969
318-339		UTD281-300H	Bristol LHL	N Counties	B39D	1970
338-339	347-348	WTD671-672H	Seddon Pennine RU	Plaxton	B40D	1970
340		WTD673H	Seddon Pennine RU	Plaxton	B40D	1970
341	349	WTD674H	Seddon Pennine RU	Plaxton	B40D	1970
342	354	WTD675H	Seddon Pennine RU	Plaxton	B40D	1970
343-349		WTD676-682H	Seddon Pennine RU	Plaxton	B40D	1970
350	350	WTD683H	Seddon Pennine RU	Plaxton	B40D	1970
351		WTD684H	Seddon Pennine RU	Plaxton	B40D	1970
352-353	352-353	WTD685-686H	Seddon Pennine RU	Plaxton	B40D	1970
354		WTD687H	Seddon Pennine RU	Plaxton	B40D	1970
355-356	355-356	WTD688-689H	Seddon Pennine RU	N Counties	B40D	1970
357		WTD690H	Seddon Pennine RU	Plaxton	B40D	1970
358-363	2358-2363	ATD272-277J	Daimler Fleetline CRG6LXB	Plaxton	H49/27D	1971
364-371	364-371	DTC712-719J	Seddon Pennine RU	N Counties	B40D	1971
372		DTC720J	Seddon Pennine RU	Plaxton	B40D	1971
373-393	373-393	DTC721-741J	Seddon Pennine RU	Plaxton	B40D	1971
394-403	2394-2403	RTJ422-431L	Daimler Fleetline CRG6LXB	N Counties	H49/27D†	1972
404-413	2404-2413	VTC494-503M	Daimler Fleetline CRG6LXB	N Counties	H49/27D	1974
414-418	414-418	TTB445-449M	Bristol RESL6G	Plaxton	DP41F	1974
419	419	VTC733M	Bristol RESL6G	Plaxton	DP41F	1974
420-423	420-423	TTB451-454M	Leyland Leopard PSU3C/2R	Plaxton	DP44F	1974
424	424	GBN331N	Leyland Leopard PSU4B/3R	Plaxton	C43F	1974
425-427	30-32	XTB750-752N	Leyland Leopard PSU3B/4R	Plaxton	C43F	1974
428-429	33-34	XTB748-749N	Leyland Leopard PSU3C/4R	Plaxton	C51F	1974
430-431	430-431	JDK921-922P	Leyland Leopard PSU3C/4R	Plaxton	B44D	1975
432-434	432-434	JDK923-925P	Leyland Leopard PSU4C/4R	Plaxton	B44F	1975

* The body on 40 was new in 1965.
† 394/8-400/3 were H47/32F.
2413 was rebodied by Northern Counties in 1983.

318		2318	Daimler Fleetline CRL6	Park Royal	H44/27F	1972
319-320		MLK597L	Daimler Fleetline CRL6	Park Royal	H44/27F	1972
319-320		2319-2320	Daimler Fleetline CRL6	Park Royal	H44/27F	1972
321		MLK584/8L	Daimler Fleetline CRL6	Park Royal	H44/27F	1972
321		2321				
322-323		MLK591L	Daimler Fleetline CRL6	Park Royal	H44/27F	1972
322-323		2322-2323	MLK603/15L			
324		2324	MLK622L	Daimler Fleetline CRL6	Park Royal	H44/27F	1972
325-326		2325-2326	MLK634-635L	Daimler Fleetline CRL6	Park Royal	H44/27F	1972
327		2327	MLH407L	Daimler Fleetline CRL6	Park Royal	H44/27F	1972
328-329		2328-2329	MLH632/75L	Daimler Fleetline CRL6	Metro-Cammell	H44/27F	1972
330		2330	MLH489L	Daimler Fleetline CRL6	Park Royal	H44/27F	1972
331		2331	TGX710M	Daimler Fleetline CRL6	Metro-Cammell	H44/27F	1973
332-333		2332-2333	MLH452/8L	Daimler Fleetline CRL6	Metro-Cammell	H44/27F	1973
334-335		2334-2335	MLH460/2L	Daimler Fleetline CRL6	Metro-Cammell	H44/27F	1973
336-337		2336-2337	MLH465-466M	Daimler Fleetline CRL6	Metro-Cammell	H44/27F	1973

318-337 were ex-London Transport DMS597, 584/8, 591, 603, 615, 622, 634/5, 1407, 632, 675, 1489, 710, 1452/8, 1460/2/5/6 in 1980.

Godfrey Abbott Group

Fleet acquired by GMT

LFS457	Leyland Titan PD2/20	Metro-Cammell	H33/29R	1954
NCK352	Leyland Atlantean PDR1/1	Metro-Cammell	H44/34F	1959
661KTJ	Leyland Atlantean PDR1/1	Weymann	L39/33F	1959
JHF821	Leyland Atlantean PDR1/1	Metro-Cammell	H43/33F	1961
3039WY	Bedford SB1	Plaxton	C41F	1961
165/7EMJ	Albion Lowlander LR7	East Lancs	H35/30F	1962
JUV538D	Bedford VAM14	Duple	C45F	1966
UTU427E	Bedford VAM14	Duple	C45F	1967
LFV676F	Bedford VAM14	Duple	C45F	1968
LLG340G	Bristol LH6L	Plaxton	C53F	1969
VMM698G	Bedford VAM70	Duple	C45F	1969
PLG316H	Bedford VAM70	Duple	C45F	1970
WMA741-3J	AEC Reliance 6U3ZR	Duple	C45F	1971
NMB275-80L	Bedford YRQ	Plaxton	C50F	1973
PTU904L	Bedford YRQ	Duple	C53F	1973
TUY381L	Bedford YRT	Plaxton	C37FT	1973
PTF629M	Bedford YRT	Deansgate	B17F	1974
YMB965M	Bedford CF	Duple	C49F*	1975
HDT621-5N	Bedford YRT	Deansgate	B17F	1975
CMA404-9N	Bedford CF	Deansgate	B17F	1975
JJA173N	Bedford CF	Deansgate	B17F	1975
JUV538D	Bedford CF			
MJA900/1P	AEC Reliance 6U3ZR	Duple	C45FT*	1975

*HDT625N was C53F; MJA900/1P were later C41FT.
LFS457 was ex-Merseyside PTE. It had been new to Wallasey Corporation.
NCK352 was ex-Ribble.
661KTJ was ex-Ribble. It was originally a Leyland demonstrator and was then owned by Bamber Bridge Motor Services.
JHF821 was ex-United Counties. They had been new to Luton Corporation.
165/7EMJ were ex-Grey Green, London.
UTU427E, VMM698G, PLG316G were taken over with the business of Pride of Sale Motor Coaches.
LFV676F, LLG340G, PTF629M were taken over with the business of Lingley's Sale-Away Touring Co of Stretford.
TUY381L was ex-Don Everall of Wolverhampton.
All others were new to Godfrey Abbott.

LUT: Vehicles added under GMT ownership

435-444	435-444	LTE486-495P	Leyland Leopard PSU3D/2R	Plaxton	B48F	1976
445-464	445-464	MTE13-32R	Leyland Leopard PSU3D/2R	Plaxton	B48F	1976
465-479	206-220	NEN952-966R	Leyland National 11351A/1R		B49F	1977
480-484	35-39	OTD824-828R	Leyland Leopard PSU3E/4R	Plaxton	C51F	1977
485-494	6901-6910	OBN502-511R	Daimler Fleetline FE30AGR	N Counties	H43/32F	1977-8
495-514	6911-6930	PTD639-658S	Daimler Fleetline FE30AGR	N Counties	H43/32F	1978-9
515-529	6931-6945	TWH690-704T	Daimler Fleetline FE30AGR	N Counties	H43/32F	1978
530-536	221-227	PTD667-673S	Leyland National 11351A/1R		B49F	1978
537-541	40-44	TWH685-689T	Leyland Leopard PSU3E/4R	Plaxton	C51F	1978
542	6946	WWH94T	Daimler Fleetline FE30AGR	N Counties	H43/32F	1979
543-565	228-250	WBN462-484T	Leyland National 11351A/1R		B49F	1979
566-569	45-48	YBN629-632V	Leyland Leopard PSU3E/4R	Plaxton	C51F	1979
570-579	6947-6956	YTE584-593V	Daimler Fleetline FE30AGR	N Counties	H43/32F	1979
580-588	6957-6965	BCB610-618V	Daimler Fleetline FE30AGR	N Counties	H43/32F	1980
589-613	6966-6990	DWH682-706W	Daimler Fleetline FE30AGR	N Counties	H43/32F	1980
614-616	49-51	DEN245-247W	Volvo B58	Plaxton	C55F	1980

6912 (496) was rebodied by Northern Counties in 1983.
6938 (522) was rebuilt as a single-decker by Northern Counties in 1986 and was renumbered 1697.

Coaches

Fleet	Registration	Chassis	Body	Seating	Year
201/3/5	GNB516-518D	Bedford VAL14	Plaxton	C47F*	1966
202/4/6	GND111-113E	Bedford VAL14	Plaxton	C47F	1967
207-212	JND207-212F	Bedford VAL70	Plaxton	C52F	1968
213-214	MND213-214G	Bedford VAL70	Plaxton	C52F	1969
215	TRJ101	AEC Reliance 2MU3RV	Weymann	C26F	1962
216	NXJ376H	Bedford VAS5	Plaxton	C29F	1970
217-224	OND734-741H	Seddon Pennine IV	Plaxton	C51F	1970
225-226	PVR225-226J	Bedford VAS1	Plaxton	C29F	1971
227-234	RNA695-702J	Seddon Pennine IV	Plaxton	C51F*	1971
235-240	TNB436-441K	Bedford YRQ	Duple	C45F	1972
241-245	TNB442-446K	Bedford VAS5	Duple	C29F	1972
246-247	TNB447-448K	Seddon Pennine T6	Duple	C51F	1972
248-249	TXJ535-536K	Bedford YRQ	Duple	C45F	1972
250-251	TXJ537-538K	Leyland Leopard PSU5	Plaxton	C53F	1972
252-255	TXJ539-542K	Leyland Leopard PSU3	Duple	C49F	1972

201-214 were Manchester 201-214.
215 was Salford 101.
216 was delivered with registration KNK375H.
*203 was C45F; 234 was C47F.
207, 211, 212 were transferred to Selnec Cheshire as 207A, 211A, 212A in 1972.
213/4/6, 225-229, 231-255 were renumbered 13, 14, 16, 25-29, 31-55 in a new coach series in 1973.

| 201 | VXJ737L | Ford Transit (petrol) | Deansgate | C12F | 1972 |

201 was renumbered 215 in 1973 and then 15 in the new coach series in 1974.

13-14	MND213-214G	Bedford VAL70	Plaxton	C52F	1969
15	VXJ737L	Ford Transit (petrol)	Deansgate	C12F	1972
16	NXJ376H	Bedford VAS1	Plaxton	C29F	1970
25-26	PVR225-226J	Bedford VAS1	Plaxton	C29F	1971
27-29	RNA695-697J	Seddon Pennine IV	Plaxton	C51F	1971
31-34	RNA699-702J	Seddon Pennine IV	Plaxton	C51F	1971
35-40	TNB436-441K	Bedford YRQ	Duple	C45F	1972
41-45	TNB442-446K	Bedford VAS5	Duple	C29F	1972
46-47	TNB447-448K	Seddon Pennine T6	Duple	C51F	1972
48-49	TXJ535-536K	Bedford YRQ	Duple	C45F	1972
50-51	TXJ537-538K	Leyland Leopard PSU5A	Plaxton	C53F	1972
52-55	TXJ539-542K	Leyland Leopard PSU3B	Duple	C49F	1972

13-16, 25-29, 31-55 were renumbered from 213-216, 225-229, 231-255.

56-61	AJA357-362L	Leyland Leopard PSU3B/4R	ECW	C47F	1973
62-65	AWH62-65L	Leyland Leopard PSU3B/4R	ECW	C47F	1973
66-68	XNA404-406L	Leyland Leopard PSU3B/4R	Duple Dominant	C49F	1973
69-75	XNE881-887L	Leyland Leopard PSU3B/4R	Plaxton Elite Express III	C49F	1973
76-79	YNA397-400M	Leyland Leopard PSU3B/4R	Duple Dominant	C51F*	1974
80-85	HNE640-645N	Leyland Leopard PSU3B/4R	ECW	C49F	1975
86-91	KDB673-678P	Bristol LHS6L	Duple Dominant	C29F*	1975
92-93	JND992-993N	Leyland Leopard PSU3C/4R	Duple Dominant	C51F*	1975
94-95	JND994-995N	Leyland Leopard PSU3C/4R	Duple Dominant	C55F*	1975
96-97	JND996-997N	Leyland Leopard PSU5A/4R	Duple Dominant	C30F	1976
98	PNE360R	Bristol LHS6L	Duple Dominant		

*79, 94 were C44F; 93 was C31F; 96 was C48F.

1-2	ANA1-2T	Leyland Leopard PSU5C/4R	Duple Dominant II	C55F*	1978
3-4	ANA3-4T	Leyland Leopard PSU3E/4R	Duple Dominant II	C51F*	1978
5-9	ANA5-9T		Plaxton Supreme	C53F	1978
16-17	RVU536-537R	AEC Reliance 6U3ZR	Duple Dominant	C44FT*	1977

18	RVU539R	AEC Reliance 6U3ZR	Duple Dominant	C44FT	1977
19-20	MJA900-901P	AEC Reliance 6U3ZR	Duple Dominant	C45FT	1975
21-24	HDT621-624N	Bedford YRT	Duple Dominant	C49F	1975
25	PTU904L	AEC Reliance 6U3ZR	Plaxton	C50F	1973
26-27	MNC500-501W	Leyland Leopard PSU5D/5R	Duple Dominant II	C42F*	1980
28-29	MNC502-503W	Leyland Leopard PSU3E/4R	Duple Dominant II	C44F	1980

26-29 were to have been registered GBA26-29V.
5-9, 16-25 were former Godfrey Abbott coaches, renumbered in 1979.
*2 was C48F; 3 was C44F; 16 was C40FT; 26 was C32FT.

30-32	XTB750-752N	Leyland Leopard PSU4B/3R	Plaxton Elite	C43F	1974
33-34	XTB748-749N	Leyland Leopard PSU3B/4R	Plaxton Elite	C51F	1974
35-39	OTD824-828R	Leyland Leopard PSU3E/4R	Plaxton Supreme	C51F	1977
40-44	TWH685-689T	Leyland Leopard PSU3E/4R	Plaxton Supreme	C51F	1978
45-48	YBN629-632V	Leyland Leopard PSU3E/4R	Plaxton Supreme IV	C51F	1979
49-51	DEN245-247W	Volvo B58	Plaxton Supreme IV	C55F	1980

30-51 were former Lancashire United 425-429, 480-484, 537-541, 566-569, 614-616. Of these, 30-34 did not receive their GMT fleet numbers.

21	BJA856Y	Leyland Tiger TRCTL11/3R	Duple Goldliner	C46F	1982
22-24	FWH22-24Y	Leyland Tiger TRCTL11/3R	Duple Goldliner	C46FT	1983
25	WBV540Y	Leyland Leopard PSU3B/4R	Plaxton Paramount 3500	C49FT	1982
30	515VTB	Setra S215HD	Setra	C49FT	1983
31	583TD	Setra S215HD	Setra	C49FT	1983
32-33	A32-33KBA	Leyland Tiger TRCTL11/3R	Duple Laser	C55F	1983
52-53	ANA52-53Y	Leyland Tiger TRCTL11/3R	Plaxton Paramount 3200	C53F	1983
54	A54KVM	Leyland Tiger TRCTL11/3R	Plaxton Paramount 3200	C55F	1984
55	A155KVM	Leyland Leopard PSU3B/4R	Plaxton Paramount 3200	C55F	1984
64	KGS491Y	Leyland Tiger TRCTL11/3R	Plaxton Paramount 3200	C53F	1983
65	KGS493Y	Leyland Tiger TRCTL11/3R	Plaxton Paramount 3200	C53F	1983
66-67	A66-67KVM	Leyland Tiger TRCTL11/3R	Plaxton Paramount 3200	C55F	1984
82-88	SND82-88X	Leyland Leopard PSU3B/4R	Duple Dominant	C51F	1975
100	476CEL		Duple Dominant	C34FL	1975

25 was an ex-Leyland demonstrator, purchased in 1984.
30-31 had registrations transferred from LUT Guy Arabs.
64-65 were ex-Londoners, London in 1985.
82-88 were rebodied by Duple in 1981-82 on the chassis of 83-89 (HNE643-5N, KDB673-6P).
100 was rebodied by Duple on the chassis of 82 (HNE642N); it ran briefly as KDB676P and was fitted to carry wheelchair-bound passengers.

1-4	B368-371VBA	Leyland Tiger TRCTL11/3R	Plaxton Paramount 3200	C53F*	1985
5	C167ANA	Volvo B10M	Plaxton Paramount 3200	C53F	1985
6-7	C706-707END	Volvo B10M	Plaxton Paramount 3200ls	C53FL	1986
8-12	C308-312ENA	Leyland Tiger TRCTL11/3RH	Duple 320	C57F	1986
13	not used				
14	C106AFX	Volvo B10M	Plaxton Paramount 3500	C49F	1986
15-16	E574/8UHS	Volvo B10M	Plaxton Paramount 3500	C53F	1988
17	E853JVR	Leyland Tiger TRCTL11/3R	Plaxton Paramount 3500	C49F	1989
18-19	F33/5HGG	Volvo B10M	Plaxton Paramount 3200	C53F	1989
20	H170DVL	Setra S215HD	Setra	C53F	1990
21	D639MDB	MCW Metrorider	MCW	C19F	1987
22	G181VBB	Setra S215HR	Setra	C49F	1990
23	G715TTY	Setra S215HR	Setra	C49F	1990

* 4 has a Plaxton Paramount 3500 C49FT body.
6, 7 have wheelchair lifts.
14 was new to Excelsior of Bournemouth and was acquired in 1989.
15, 16, 18, 19 were new to Park of Hamilton and were acquired in 1989 (15/6) and 1990 (18/9).
20 was registered by the supplying dealer, Kassbohrer UK.
21 was reseated and renumbered from 1639 in 1990.
22/3 were ex Craigs of Amble in 1991.

500 series: Leyland Lynxes

501-504	D501-504LNA	Leyland Lynx	Leyland	B48F	1986

700 series: Volvo B10M

701	J461OVU	Volvo B10M	Northern Counties	B49F	1991

2000 series: Dennis Dominators

2001-2020	B901-920TVR	Dennis Dominator	Northern Counties	H43/32F	1985
2021-2030	B21-30TVU	Dennis Dominator	Northern Counties	H43/32F	1985
2031-2040	H131-140GVM	Dennis Dominator	Northern Counties	H43/29F	1991

3000 series: Leyland Olympians

3001-3010	ANA1-10Y	Leyland Olympian ONTL11/1R	Northern Counties	H43/30F	1982-83
3011-3015	A576-580HDB	Leyland Olympian ONTL11/1R	Northern Counties	H43/30F	1983
3016	A581HDB	Leyland Olympian ONLXB/1R	Northern Counties	H43/30F	1983
3017-3020	A582-585HDB	Leyland Olympian ONLXCT/1R	Northern Counties	H43/30F	1983-84
3021-3025	A21-25HNC	Leyland Olympian ONLXCT/1R	Northern Counties	H43/30F	1984
3026-3033	A26-33ORJ	Leyland Olympian ONLXBA/1R	Northern Counties	H43/30F	1984
3034-3035	B34-35PJA	Leyland Olympian ONLXBA/1R	Northern Counties	H43/30F	1984
3036-3049	B36-49PJA	Leyland Olympian ONLXB/1R	Northern Counties	H43/30F	1984
3050-3051	B350-351PJA	Leyland Olympian ONLXB/1R	Northern Counties	H43/30F	1984
3052-3085	B52-85PJA	Leyland Olympian ONLXB/1R	Northern Counties	H43/30F	1984-85
3086-3115	B86-115SJA	Leyland Olympian ONLXB/1R	Northern Counties	H43/30F	1984-85
3116-3125	B116-125TVU	Leyland Olympian ONLXB/1R	Northern Counties	H43/30F	1985
3126-3145	B126-145WNB	Leyland Olympian ONLXB/1R	Northern Counties	H43/30F*	1985
3146-3155	B146-155XNA	Leyland Olympian ONLXB/1R	Northern Counties	H43/30F	1985
3156-3200	C156-200YBA	Leyland Olympian ONLXB/1R	Northern Counties	H43/30F*	1985-86
3201-3225	C201-225CBV	Leyland Olympian ONLXB/1R	Northern Counties	H43/30F	1986
3226-3235	C226-235ENE	Leyland Olympian ONLXB/1R	Northern Counties	H43/30F	1986
3236-3238	C236-238EVU	Leyland Olympian ONLXB/1R	Northern Counties	H43/30F	1986
3239-3245	C239-245EVU	Leyland Olympian ONLXB/1R	Northern Counties	CH43/26F	1986
3246-3255	C246-255FRJ	Leyland Olympian ONLXB/1R	Northern Counties	CH43/26F	1986
3256-3277	D256-277JVR	Leyland Olympian ONLXB/1R	Northern Counties	CH43/26F	1986
3278-3305	F278-305DRJ	Leyland Olympian ONLXB/1R	Northern Counties	H43/30F	1988-89

* 3139, 3198, 3213/4 converted to CH43/16F for Airport Express.
3001-3025 were to have been ANA1-25Y.

4000 series: Leyland Titans

4001-4002	ANE1-2T	Leyland Titan TNLXB/1RF	Park Royal	H46/27F*	1978-79
4003	FVR3V	Leyland Titan TNTL11/1RF	Park Royal	H46/27F	1979
4004-4005	ANE4-5T	Leyland Titan TNLXB/1RF	Park Royal	H46/27F*	1979
4006-4011	GNF6-11V	Leyland Titan TNLXB/1RF	Park Royal	H46/27F*	1979-80
4012-4015	GNF12-15V	Leyland Titan TNTL11/1RF	Park Royal	H46/27F	1980

*4001/6/7 were H47/26F
Registrations and fleet numbers were allocated to 35 Titans:
4001-4035 were to have been ANE1-35T. When deliveries were delayed 4006-4025 were allocated registrations GNF6-25V. GNF16/7V were issued to Standard Atlanteans 8141/2.

5000 Series: MCW Metrobuses

5001-5010	GBU1-10V	MCW Metrobus DR101/6	MCW	H43/30F	1979
5011-5017	GBU11-17V	MCW Metrobus DR102/10	MCW	H43/30F	1980
5018-5019	MNC494-495W	MCW Metrobus DR102/10	MCW	H43/30F	1980
5020	GBU20V	MCW Metrobus DR102/10	MCW	H43/30F	1980
5021	MNC496W	MCW Metrobus DR102/10	MCW	H43/30F	1980
5022	GBU22V	MCW Metrobus DR102/10	MCW	H43/30F	1980
5023	MNC497W	MCW Metrobus DR102/10	MCW	H43/30F	1980
5024-5025	GBU24-25V	MCW Metrobus DR102/10	MCW	H43/30F	1980
5026	MNC498W	MCW Metrobus DR102/10	MCW	H43/30F	1980
5027-5029	GBU27-29V	MCW Metrobus DR102/10	MCW	H43/30F	1980
5030	MNC499W	MCW Metrobus DR102/10	MCW	H43/30F	1980
5031-5070	MRJ31-70W	MCW Metrobus DR102/21	MCW	H43/30F	1980-81
5071-5100	ORJ71-100W	MCW Metrobus DR102/21	MCW	H43/30F	1981
5101-5110	SND101-110X	MCW Metrobus DR102/21	MCW	H43/30F	1981
5111-5150	SND111-150X	MCW Metrobus DR102/23	MCW	H43/30F	1981-82
5151-5190	ANA151-190Y	MCW Metrobus DR102/23	MCW	H43/30F	1982-83
5201-5210	C201-210FVU	MCW Metrobus DR132/8	Northern Counties	CH43/29F	1986
5301-5310	D301-310JVU	MCW Metrobus DR102/51	Northern Counties	CH43/29F	1986
5311-5320	D311-320LNB	MCW Metrobus DR102/51	Northern Counties	CH43/29F	1987

5001-5030 were to have been registered ABA1-30T, then GBU1-30V. Registrations ABA11-30T were used for Leyland Nationals 186-205.

7000 series: Volvo Citybuses

7001-7008	H701-708GVM	Volvo Citybus	Northern Counties	H45/32F	1991
7009-7010	J709-710ONF	Volvo Citybus	Northern Counties	H45/28F	1992

7009/10 have wheelchair lifts; they were to have been H709/10GVM.

Appendix 1: The Fleet at Privatisation

Double-Deckers	GMBN	GMBS	Total	Single-Deckers	GMBN	GMBS	Total
Standard Atlantean	427	297	724	Dodge S56/Renault S75	96	56	152
Leyland Olympian	165	141	306	MCW Metrorider	11	74	85
MCW Metrobus	118	99	217	Dennis Domino	–	20	20
Standard Fleetline	104	21	125	Leyland Leopard	19	–	19
Dennis Dominator	–	40	40	Leyland National	7	8	15
Volvo Citybus	13	–	13	Leyland Lynx	4	–	4
Scania	–	7	7	Volvo Citybus	1	–	1
Dennis Falcon	–	3	3	Coaches	–	29	29
Total	827	608	1,435	Total	138	187	325

GM Buses North, fleet at privatisation:

Leyland National	132, 179, 197, 201-3/47
Leyland Leopard	436/8-42/4, 451/2/4-63
Leyland Lynx	501-4
Volvo Citybus single-deck	701
Volvo Citybus double-deck	1481-3, 7001-10
MCW Metrorider	1616/22/5/31/81-7
Renault S75	1722/3/5
Dodge S56	1777/8/80/3-6/90/4-6/9, 1845-7/50/2/4/7/64/70/1-4-6/9/85/9/94/7/9, 1904/6-10/4/6/8-20/5/7-36/40-3/5-50/2/4/6/8/60/1/5-7/9/70/2/3/5-80/3-7/9/91/4-7
Leyland Olympian	3011-5/37/8/40-8/50-2/4/9/61-4/6/8/71/3/5/6/8/9/81/3/5/90/2/3/6-9, 3100-9/11-3/5/6/20/3/7-31/4/6/40-2/4/8/51/2/7/9-63/8/71/7/80/2/3/6-90/2/4, 3200-4/6/9/11/7-20/2/3/5/7-9/31-3/5/7-54/6-9/61-7/70/1/3-6/8-81/4/6-8/90/2/3/9, 3302/3/5
Leyland Fleetline	4026/39/46/50/1/4-6-8/60/2-4/72-5/7-9/81/4/6/8/90/1/3/5/6, 4102/4-6/8-11/3-5/7-23/5-33/6/9-43/6-50, 4914/42-9/52/3/5/6/61-5/7/9/70-7/9/81/2/5-90
Leyland Atlantean	4151/4/5/8-60/3/5/7/71/2/7/9-81/3/4/9/91/2/4/6-9, 4201/2/5-9/13/6/9/21/3/4/6/8/9/32/3/8-40/3/4/51/3-8/60-3/5-9/71-9/82/4-9/92/3/6-9, 4300/3/5/8/10/1/3/5/6/8/20/2-4/6/30/1/3-8/40-2/4-7/9-51/3-61/3/4/6/7/9-74/8/9/82-92/4/8/9, 4400/1/3/5/7/8/10/1/3-7/9/20/3/4-6-8/31/3/4/6/8/9/41-8/56-61/3/4/6/9-71/4/5/8/80-2/5/6/90-4/7-9, 4500/2-4/7-11/5-7/20/2-5/8/9/31/2/4-6/9-42/4/7-9/51/4-8/60-3/6/7/70/1/3-6/8/80/1/3/4/7/8/90/1/4/5/8/9, 4602/3/6/7/10/1/4-9/21-3/6/8/9/33-6/8/40-3/5/8-50/2/4-6/8/9/62/3/6/7/70/2/3/6/7/81/2/5/6/9/91/2/7, 4701/3/7/9/12/3/6-8/20/1/3/4/7-9/31-3/6-40/2/6/53/5/6/8/60/3/5, 7077, 7685/99, 7719/21/4-6/60/2/7-9/71/3/9/80/2-5/7/91/3, 7802/3/5/6/9/21/2/8-30/3-5/7/9/46/7/52/4/66/7/83/6/98, 7902/4/11/4/21/2/8/33/4/7/49/51
MCW Metrobus	5011-5/31-5/9/50/6-64/8-70/82-90/6/7/9, 5101-5/12-5/22/6-31/3-40/6-52/6/66-9/71/2/4-8/81-9, 5201-10, 5301-20

GM Buses South, fleet at privatisation:

Coaches	1-12, 14-20, 22/3, 30/2/3, 64/5, 82/6, 100
Leyland National	160, 190/2/4/9, 205/32/40
Leyland Olympian	1451, 3001-10/6-36/9/49/53/5-8/60/5/7/9/70/2/4/7/80/2/4-6-9/91/4/5, 3110/4/7-9/21/2/4-6/32/3/5/7-9/43/5-7/9/50/3-6-8/64-7/9/70/2-6/8/9/81/4/5/91/3/5-9, 3205/7/8/10/12-6/21/4/6/30/4/6/55/60/8/9/72/7/82/3/5/9/91/4-8, 3300/1/4
Scania	1461-7
Dennis Falcon	1471-3
MCW Metrorider	1601-15/7-21/3/4/6-30/2-6/8-68/70-80
Renault S75	1721/4/6-8
Dennis Domino	1751-70
Dodge S56	1759/63/72-6, 1781/2/7-9/91-3/7/8, 1800/61/5/91/2/6/8, 1911-3/7/21-3/6/38/44/51/3/7/9/62-4/8/71/4/81/8/90/2/3/8-2000
Dennis Dominator	2001-40
Leyland Fleetline	4002/20/1/4/35/7/42/3/61/8/98, 4100/1/3/7/16/35/7/8/44/5
Leyland Atlantean	4152/3/6/7/62/4/6/8-70/4/5/8/82/6-8/90/3/5, 4200/3/4/10/2/4/5/7/8/20/2/5/7/30/1/4-7/41/2/5-7/9/50/2/9/64/80/1/3/90/1/4/5, 4301/2/4/7/9/12/4/7/9/21/5/7-9/32/9/43/8/52/62/5/8/75-7/80/1/93/5/6, 4402/4/6/9/12/8/21/2/5/9/30/2/5/7/40/9-55/62/5/8/72/3/6/7/9/83/4/7-9/95/6, 4501/5/6/12-4/8/9/21/6/7/3/0/7/8/43/5/6/50/2/3/9/64/8/9/72/7/9/82/5/6/9/92/3/6/7, 4600/1/4/5/8/9/12/3/20/4/5/7/30-2/7/9/44/6/7/51/3/7/60/1/4/5/8/9/71/4/5/8-80/3/4/7/8/90/3-6/8/9, 4700/2/4-6/8/10/1/4/5/9/22/5/6/30/4/5/41/3-5/7-52/4/7/9/61/2/4, 7032, 7564, 7687/96, 7701/18/29/36/59/66/78, 7813-5/20/31/8/43/5/56/60/2/8/70/2/8/9/81/5/9/91/2/4-6/9, 7900/3/8/12/3/5/9/24-6/9/39/41/4/5/8/53-6/8-60
MCW Metrobus	5017-24/6-30/6-8/40-9/51-5/66/7/71-81/91-5/8, 5100/6-11/6-21/3-5/32/41-5/53-5/7-65/70/3/9/80/90

Appendix 2: Key PTE statistics, 1970-1986

Year	Fleet	Staff	Passengers (millions)	Revenue (£000)	Operating Costs (£000)
1970	2,439	10,252	623	23,134	22,138
1971	2,323	10,054	476	21,483	21,582
1972	2,610	10,733	523	23,836	25,115
1973	2,601	10,413	524	24,950	28,296
1974/75	2,804	11,502	677	35,045	52,825
1975/76	3,065	12,500	483	40,804	56,100
1976/77	3,026	12,056	455	49,510	65,669
1977/78	2,987	12,181	436	54,296	75,821
1978/79	2,930	12,197	418	59,365	82,891
1979/80	2,882	11,720	411	70,246	100,693
1980/81	2,813	10,682	371	78,275	113,990
1981/82	2,590	9,806	353	78,309	120,577
1982/83	2,530	9,730	342	82,487	126,861
1983/84	2,486	9,474	342	80,000	134,540
1984/85	2,432	9,144	346	81,384	138,050
1985/86	2,370	8,401	350	81,322	147,210

NOTE

The figures for 1970 cover 14 months to 31st December. The figures for 1974/75 cover 15 months from 1st January 1974 to 31st March 1975. Revenue figures between 1973 and 1975/76 include revenue from other operators running within the PTA area. Changes in methods of calculation may render year-on-year comparisons invalid – figures are purely a guide to key statistics.